Water Balances in the Eastern Mediterranean

edited by
David B. Brooks and Ozay Mehmet

INTERNATIONAL DEVELOPMENT RESEARCH CENTRE
Ottawa • Cairo • Dakar • Johannesburg • Montevideo • Nairobi • New Delhi • Singapore

Published by the International Development Research Centre
PO Box 8500, Ottawa, ON, Canada K1G 3H9

January 2000

Legal deposit: 1st quarter 2000
National Library of Canada
ISBN 0-88936-907-0

The catalogue of IDRC Books may be consulted online at http://www.idrc.ca/index_e.html

This book may be consulted online at http://www.idrc.ca/books/focus/907

CONTENTS

Acknowledgments

The editors would like to acknowledge the cooperation of the International Development Research Centre (IDRC) and Carleton University in making their facilities available for the workshop. Two groups contributed the funds that made the workshop possible; first, the Expert and Advisory Services Fund in support of the Middle East Peace Process, a fund that is provided by the Canadian International Development Agency and the Canadian federal Department of Foreign Affairs and International Trade and that is managed by IDRC; second, the Faculty of Public Affairs and Management at Carleton University. Their joint support was essential to the international collaboration that we were able to effect at this workshop.

We must also acknowledge the assistance of a number of people who contributed significantly to the workshop and to the book: Margaret Emokor and David Todd of IDRC, together with Habib Mehiri, helped enormously with plans for an open session presented to journalists, government officials, representatives from the diplomatic community, and members of nongovernmental organizations. Ms Emokor also assisted with the administration of the IDRC grant to Carleton University that was essential to making the workshop possible. Ms Keri Holtby, a graduate student at the Norman Patterson School of International Affairs, served as rapporteur at the workshop, and Michael Benedeck and Ann-Lisa Brewer, also graduate students, assisted with the practical aspects of the workshop. Ms Enid Valerie provided literary and production editing of each of the chapters. We are enormously grateful for the help of all of these people.

Executive Summary

Ozay Mehmet

Nature and scope of the workshop

The papers published in this book were originally presented at a one-and-one-half-day workshop held at the International Development Research Centre (IDRC) and Carleton University, both in Ottawa, Canada, on 29–30 October 1998. The workshop focused on freshwater balances in the Eastern Mediterranean region. The choice of water balances reflects the view of the editors that this subject represents the critical first step in exploring prospects for regional cooperation in the sharing and management of fresh water. The decision to focus on the Eastern Mediterranean — territorially defined as the arc of nations bordering on the eastern Mediterranean Sea from Turkey in the north to Egypt in the south — reflects the fact that even as fresh water is rapidly becoming a global problem, it is in the Eastern Mediterranean where the situation is closest to reaching crisis proportions. People living in this part of the world have only one-sixth as much fresh water available per capita as the global average.

The design of the workshop was unique in two respects. First, the nature of the workshop was academic: invited speakers participated and shared their views in terms of their multidisciplinary research, without any political or diplomatic representation. Second, the geographic scope of the workshop was deliberately based on a broad definition of the Eastern Mediterranean region, and the purpose of this was to transcend the definition of the Middle East freshwater issue as strictly an Arab–Israeli problem. This perspective has proven inadequate for solving the long-term Middle East water shortage because it implies a zero–sum view of the problem. One option for breaking out of the zero–sum dilemma would be to, for example, import fresh water on a large scale. Such imports, perhaps from Turkey, could serve as a basis of wider regional cooperation. With this in mind, Turkish researchers were invited to participate in the workshop, and everyone was asked to examine, among other alternatives, the costs and benefits of large-scale water imports. Basically, water imports imply a new, international water market. Such a market is slowly emerging in the region, starting on a small scale with shipment of Turkish water to North Cyprus, a topic of one of the papers presented at the workshop (Biçak and Jenkins, this volume), and more broadly representative of another way to break out of the zero–sum dilemma.

One major omission of the workshop was Syrian input. Although efforts were made at the planning stage to include representation from Syria, in the end this proved infeasible; but it is hoped that this deficiency will be remedied at future workshops. A new development that reflects our changing political times was that the West Bank and Gaza were treated by all participants as an entity with sovereign rights under the name of Palestine.

Understanding water balances

Determining the water balance for a country or a region is a complex undertaking because of the great variety of factors that influence water supply and demand. On the supply side, weather conditions cause variations from season to season and year to year; and, in the longer term, climatic change affects precipitation and evaporation over large areas. On the demand side, population, agriculture, and industry, along with pricing and subsidy levels for all three factors, greatly influence consumption and the economic use of water. Examining both the supply and demand sides of the equation together, scientific estimates of water balances depend on bureaucratic efficiency, particularly the assiduousness and accuracy of statistical data collection. Last but not least, political factors include national boundaries on the map that do not coincide with natural watersheds and strategic security concerns among neighbouring countries that give rise to conflict between upstream and downstream states. All these difficulties exist in the Eastern Mediterranean region, indeed perhaps to a greater extent than in other parts of the world.

Despite all these difficulties, regional cooperation in the Eastern Mediterranean is essential. This is partly because of the need to consolidate and strengthen the Arab–Israeli peace process; but even more fundamentally regional cooperation in freshwater management is essential because the region as a whole is already experiencing critical water shortages. Projections indicate that unless the shortage is effectively and equitably remedied in the meantime, it will reach crisis proportions by 2025. New technologies in the field of desalinization of seawater and recycling of wastewater and the emergence of an import–export market in water offer some intriguing opportunities for solving, or at least mitigating, the broader regional water crisis in the Eastern Mediterranean early in the next century. More immediately, water demand management, including higher prices for water, must play a leading role (probably the leading role), despite the fact that per capita consumption of water in this region is already well below world averages.

The key elements of the presentations delivered at the workshop are summarized below in four main parts: first, an agenda for research on water in the Eastern Mediterranean; second, a number of country-by-country reviews of national water balances; third, a cross-cutting paper on regional cooperation, with a focus on options to promote successful negotiations; and, fourth; a resolution on a plan of action that participants at the conference adopted. It is hoped that the chapters in this volume represent a modest start to a difficult task, but an essential step on the road to wider regional cooperation in such a vital and critical development resource as fresh water.

Agenda for research on water in the Eastern Mediterranean

David B. Brooks, of IDRC, keynoted the workshop with a talk in which he presented a research agenda for working toward regional cooperation. His written presentation begins by emphasizing that after being neglected for many years, fresh water is rapidly becoming the global resource issue. Evidence of this is found in the large number of studies on the subject from a global perspective, together with growing attention to water as an economic good, as well as a natural resource with aesthetic, cultural, and social dimensions. Water is also drawing more attention in studies of inter- and intranational conflict resolution, although it is becoming clear that water is seldom if ever the prime cause of war.

Brooks went on to suggest that, from an analytical, if not a political, perspective, the situation confronting academics on the subject of water today is much like that of the OPEC crisis in October 1973. Therefore, it is appropriate to look at the immediate and medium- and longer term needs for research. In each case, he argues, those subjects simultaneously most relevant to policy and most seriously in need of study are neither technical nor narrowly economic but broadly socioeconomic, sociopolitical, and even sociopsychological. The short-term agenda includes

- Water demand management, with emphasis on why it is not currently playing a greater role in terms of regional cooperation;

- Opportunities for greater use of water that is marginal in quantity or quality, or both; and

- Special attention to the ecological demand for water.

The medium-term agenda includes

- Efficient and equitable methods for reallocating water from agricultural to other uses;

- Institutions and programs for local water management;

- Institutions and measures to deal with periods of prolonged drought; and

- Institutions to manage transboundary water supply and pollution.

The long-term agenda includes only one main proposal: a common analytical approach to be used to review the range of megaprojects suggested for the ultimate resolution of freshwater supply–demand problems in the Eastern Mediterranean region, including desalination, canals, and importation by tankers, balloons, or pipelines.

National water balances

Lebanon

Hussein A. Amery, of the Colorado School of Mines in the United States, begins his chapter by noting the diversity of estimates of Lebanon's water balances. Two such estimates are 2 600 to 3 375 Mm^3/year. He argues that many factors account for the discrepancy between these figures, including timeliness of the estimates, climatic factors, and the politics of boundaries.

Amery's chapter highlights the key hydrogeographical features in Lebanon: two mountain ranges running north–south, the Lebanon range parallel to the coast and the anti-Lebanon range parallel to the eastern border, with the Bekaa Valley between them. Because of the rain shadow created by the Lebanon range, precipitation is heavy along the coastal plains but much less so in the interior. Topographically, Lebanon is broken into four hydrological zones with three primary rivers: the Litani, the Assi–Orontes, and the Hasbani. Amery's paper concentrates on the first two.

At present, Lebanon's river systems are far from having optimal development for irrigation or for hydroelectric generation. Although there are proposals for up to 40 new dams, current plans focus on 16 to 20 of these, most of which would serve to store excess precipitation from the wet season (November to April). However, in the southern section of the Litani River, a 600-m drop over 60 km produces a water flow that could be harnessed for hydroelectric generation.

The Litani is Lebanon's largest river in terms of both "flow and passion." It "represents the life of the Lebanese." Many hold that the Litani is the key to the economic success of Lebanon. However, there is widespread, if unspoken, fear of losing Lebanon's ability to control the Bekaa Valley, the source of the Litani. One consequence of this situation is that it is hard to get accurate figures about water flows. Most records are from the 1920s and 1930s. Moreover, the civil war has resulted in a scarcity of recent water data in Lebanon; whatever data were collected were gathered haphazardly by "keen employees." The official database remains inadequate and out of date.

There are also problems with estimates of renewable water resources in Lebanon. Measurement is difficult because it is impossible at present to determine what is and is not Lebanese water. The methodology for collecting data is not made available, so it remains unclear how the ministry responsible for water has arrived at official numbers. It is not surprising, therefore, that figures on water supply and consumption are contradictory. In terms of agriculture, 87 500 ha is under irrigation, with potentially 177 500 ha available for irrigated agriculture. Urban and rural Lebanese also have an ongoing dispute over the distribution of water, including discussions for a project that would divert 500 000 Mm^3/day of water away from agriculture to urban uses in the capital, Beirut. Two projects were started in the summer of 1998 to build irrigation canals.

The Assi River originates in Lebanon, passes through Syria, and then enters the Mediterranean on the coast of the province of Hatay in Turkey. The Assi case is a component of the Turkish–Syrian dispute over water sharing. At present, however, Lebanon and Syria have a deal to share the waters of the Assi: if water flow is 400 Mm^3/year or more, Lebanon gets 80 Mm^3 for its own use.

Water quality is also emerging as a concern in Lebanon. Lebanese Greenpeace has found toxins sitting next to the Litani River on top of the aquifer. A significant factor in water quality is the inadequacy of the existing sewer infrastructure. Only 71% of population has the use of a sewer system.

In conclusion, Amery states that although Lebanon is currently rebuilding its economy and infrastructure at a rapid rate, it is neglecting concerns about water quality and quantity. Water production and distribution systems are inadequate. Consequently, Lebanon is water poor, largely because its water infrastructure is grossly inadequate. The nation needs to refurbish and expand its existing water infrastructure, update hydrological maps, and provide better drainage.

Israel

In Harvey Lithwick's comprehensive chapter on Israel, he begins with an examination of the traditional factors affecting water balances. He provides estimates of water stocks and flows from major sources up to 1990. His primary concern is the mechanism by which these supplies have been allocated and how efficient that allocation process is. Using a traditional economic analysis, he examines alternative estimates of present and future water supply and demand to arrive at its scarcity value. This analysis provides the basis for an examination of alternative water policies to influence both supply and demand, as well as providing methods to manage that process. The central role of water pricing is emphasized.

Currently, Israel gets most of its fresh water (more than 2000 Mm^3/year) from three sources: the Sea of Galilee, the coastal aquifer, and the Yarkon aquifer. In addition, it has made increasing use of recycled sewage water, amounting to 453 Mm^3/year. One third of the sewage water is treated and used for irrigation, mainly in the arid Negev region, where agricultural land in the centre of the country gives way to urban sprawl. Significant portions of Israel's freshwater supply are under dispute with its neighbours. Water losses occur as a result of evaporation and, in several urban areas, as a result of broken water pipes.

Major gains in water supply will likely arise from two sources:

- Desalination of seawater, which is promoted aggressively, on security grounds. Because the cost of desalination is now estimated at between 0.80 and 1.00 United States dollars (USD)/m^3, which is higher than the opportunity cost of current supplies, it will only become a cost-effective alternative in a decade or so. To pursue this option, capital investment must begin shortly, but once this is begun, Israel would be locked into this alternative.

- The only alternative for new large-scale supplies is water importation from Turkey, via tankers or Medusa bags. The real cost of delivering such water is very competitive, but Turkey's pricing policy is making it too expensive to deter the rapid move toward the desalination alternative.

Several other alternatives are available, although their small scale or high cost, or both, at this time make them unlikely to be serious contenders with the options of

- Capturing rainwater and storing it in microdams and ponds (a problem identified early with this solution, that of rapid silting up, is being resolved through the systematic mining of the silt for use in construction); and

- Construction of canals from either the Mediterranean or the Red Sea to the Dead Sea. (Although the water would be saline, it could be used for agriculture and would also provide sufficient fall to generate electricity and help replenish the rapidly depleting Dead Sea.)

The alternative of more efficiently using existing supplies has also been vigorously pursued. In particular, the use of water in agriculture has been improved dramatically through innovations in drip irrigation, the use of plastic hothouses, computerized fertilization, pest-control systems, and improved varieties of crops better able to survive on less water and on saline water.

A more efficient regime of water pricing would enhance these benefits. On the demand side, a move toward opportunity-cost pricing has already accelerated innovations and stimulated more suitable crop selection in the formerly heavily subsidized agricultural sector. Steeply rising marginal-cost pricing for residential use of water has had a similar restraining effect. But recent rapid population growth resulting from immigration has mitigated these gains to some extent. On the supply side, more realistic pricing has already brought forth a variety of alternative supplies, such as the use of recycled sewage water for irrigation. It should be noted, however, that this has external costs, especially those of an environmental nature, which may seriously restrict its longer term feasibility.

Even with these efficiency gains, Israel faces two major problems. One is that in the longer term, supplies will have to be enhanced from the options noted above. And better means will have to be found to deal with the periodic droughts that affect the region as a whole. To date, region-wide solutions have been difficult to achieve because of continuing conflict and mistrust. Water rights are seen as nonnegotiable elements of national security. This attitude leads to inefficient allocations and, hence, waste and shortages, which in the end aggravate the situation. Lithwick argues that rational management based on realistic pricing can provide a region-wide basis for collective solutions to both of these problems, both in the short and longer terms. As a bonus, a cooperative water policy may help advance the peace process by clarifying areas of common interest.

Palestine

Samer Alatout, who is currently completing his doctoral studies at Cornell University in the United States, presented a paper on water balances in Palestine that focuses on underground water sources from four known aquifers and emphasizes that Palestine has a claim on 1 666 to 2 045 Mm^3 of water resources but only uses about 125 Mm^3 of this amount. Compared with Israel and the United States, Palestine uses a relatively small amount of water. Going beyond the numbers, Alatout discusses four interrelated issues.

First, he critically examines the scientific legitimacy and technical authority of various estimates of Palestine's water balance. His main conclusion is that these estimates are both technically uncertain and reflective of cultural and political contexts of the region.

Second, Alatout argues that the practice of estimating water balances is deeply embedded in politics, an issue seldom considered in the literature. Using new insights from the field of science and technology studies, Alatout points out that the seemingly neutral and objective language used to discuss Palestine's water balance is, in actual fact, highly political. This political context is not referred to in the terms typically used by water experts to refer to political interests but in terms of an observable culture of water expertise in the Middle East. This is produced within, and shaped by, the institutional, cultural, and political contexts. He suggests that "the role of those looking for a workable water-sharing regime between Israel and Palestine is to unpack and make explicit the politics of technical jargon, rather than adding additional layers of obscurity" (Alatout, this volume).

Third, Alatout points out that, because most of the water resources of Palestine are shared with other parties, the issue is subject to international law. Although the international legal principle of historic right continues to be appropriate to the Palestinian–Israeli context, it is insufficient. No less significant in determining Palestine's water share is the international legal principle of the equitable use of shared water resources. Applying these principles requires constant negotiation and collaboration among Palestinians, Israelis, and international water experts and policymakers.

Finally, Alatout concludes by reclaiming the notion of water scarcity as a concept that is and should be grounded in daily lived experience, rather than in technical rhetoric.

Jordan

Esam Shannag and Yasser Al-Adwan,[1] of Yarmouk University in Irbid, Jordan, concentrate on the Jordan River basin, which many in the past have considered a potential source of conflict, but which they see as a potential source of peace-building and cooperation. Regional cooperation in water management is critical to Jordan, as all of its water originates outside the country. Accordingly, Jordan has a strong interest in systems for sharing transboundary water resources.

Jordanian water policy is currently centred on the following four priority areas:

- Demand management — The government applied a new price mechanism that increases the price of water once a certain amount is consumed. The results of this system have not yet been analyzed; however, they speculate that the middle class will be most affected by the changes.

---

[1] Shannag was not able to participate in the workshop; the paper was delivered by Al-Adwan.

- Public education and promoting water awareness — The government has used both cultural values and Islamic teachings to raise public awareness of the need for more responsible use of water.

- A major project to develop an aquifer in North Jordan — This is still in the planning stage but has an estimated cost of 600 million USD.

- Drawing water from Israel and Turkey through pipelines.

Water issues are very much in the minds of politicians and strategists in Jordan. However, it is felt that projections and scenarios do not work in the Middle East because its political dynamics defy prediction. For example, a completely unforeseen influx of refugees from the Gulf War caused the population in Jordan to increase suddenly by 10% in 1990, which created major social and economic dislocations. It would be more appropriate to be flexible and to see the "other" in terms of "us," if regional cooperation is to yield concrete results.

Regarding water security and availability, Shannag and Al-Adwan note the following:

- Security in controlling head waters is important; however, this should not be the concern of one government alone but of the entire region;

- Safety nets in water management are needed to ensure equitable distribution while meeting the demands and needs of everyone in the region;

- Schemes of cooperative regional management of water are required to promote a political and managerial resolution to the issue of scarcity; and

- A cost–benefit analysis may be appropriate if water is to be treated as an economic good, but, at the same time, other factors must be considered, such as the productive use of the work force to achieve goals for the regulation of water.

Turkey

Mehmet Tomanbay, of Gazi University, Ankara, Turkey, begins by noting that although Turkey may have the largest volume of water in the region, it is not a water-rich country. For one thing, it has only one-fifth of the water that true water-rich countries, such as Canada, have. For another, inequalities exist in water supply from region to region and from month to month within Turkey. Average yearly precipitation in Turkey is 643 mm but differs greatly across the country. For example, Istanbul and Ankara, as well as central and southern Anatolia, face water shortages, especially in the summer months. Rapid population and industrial growth and increases in the standard of living have all put stress on freshwater supply and reduced the amount of water available for domestic use and for potential export.

The Euphrates and the Tigris river basins are the two most important in terms of water flows: between them they account for 28.5% of total surface flow in Turkey. Despite growing domestic water needs, and especially in view of the massive Guneydogu Anadolu Projesi (GAP, Southeast Anatolia Project), the Turkish government is committed to sharing water equitably with its neighbours, Iraq and Syria. In addition, Turkey is willing to export the water that is now flowing freely from its national rivers into the sea. It has built a water-export terminal for this purpose at Manavgat near Antalya and is already shipping water to North Cyprus in Norwegian balloons (see Biçak and Jenkins, this volume).

GAP is a multipurpose river-basin development, largely financed from national sources, in part because of the ongoing dispute with Iraq and Syria over water sharing. Its two main objectives are

- To develop Turkey's poorest and most underdeveloped region; and

- To reduce economic disparity and raise the economic welfare of the Kurds who largely populate the region.

GAP includes 22 dams, 19 hydroelectric plants, and 2 irrigation tunnels, as well as 13 subprojects. The master plan is to transform the region into an economic and export base. After completion, early in the 21st century, crop output is expected to double or even triple; the Turkish economy as a whole, to grow by 12% or more; employment, to increase in the area; and gross regional product, to grow fourfold, with a consequent sharp decline in rural out-migration. The project, now about 40% complete, is concentrating on building infrastructure, improving health care, and developing tourism as new enterprises for the region. In recent years, greater emphasis has been on social issues, especially sustainable human development. Tomanbay believes that GAP will be finished earlier than originally expected, as a result of the government's eagerness to reap its socioeconomic benefits.

North Cyprus

Hasan Ali Biçak, of Eastern Mediterranean University in Famagusta, North Cyprus, has been collaborating, during a sabbatical leave, with Glenn Jenkins of the Harvard Institute for International Development. They review water balances in North Cyprus. They note that it has, for some time now, suffered a steady reduction in rainfall and that this problem has been aggravated through the wasteful, traditional (flooding) methods of irrigation. The result has been overextraction of water from the country's aquifers. So far, national measures to mitigate the supply-demand deficit appear to have been barely sufficient, and Cyprus as a whole is experiencing severe water shortages.

Biçak and Jenkins go on to identify the modest potential for new water resources and more significant potential for greater efficiency of water use. Key options include

- The modern irrigation project in Guzelyurt, the centre for citrus growing, where the old flooding system for irrigating citrus trees is now being replaced by a drip method that mixes water from the reservoirs with fertilizer;

- New and deeper wells;

- Wastewater treatment plants; and

- The Yesilirmak Dam project.

In the summer of 1998, North Cyprus began to import water from Turkey on a pilot-project basis. A Norwegian firm ships water from Turkey in Medusa bags towed behind ships and sells it to North Cyprus for 0.55 USD/m³. These plastic "balloons" have a capacity of 10 000 to 20 000 m³ each. This shipment technology has had some problems, such as with mooring the water bags at the receiving terminal, as well as difficulties with insurance policies and finance, but none of these problems is insurmountable.

Importation of water from the Turkish mainland to North Cyprus is small scale, but it represents a significant model for testing new technologies for large-scale water shippage to the emerging Middle East water market. The balloon technology, currently used by the Norwegian firm, is one option. Alternative technologies include tankers and pipelines. Indeed, a feasibility study conducted by Biçak and Jenkins concluded that the unit transportation cost of water imported from the Manavgat export terminal in Turkey to Kumkoy in North Cyprus would be on average 0.465 USD/m³. This compares with 0.55 USD for the Norwegian balloon alternative. Both figures simply represent shipping costs and do not include charges for the water at source. Nor do they include infrastructure or operating costs in North Cyprus. When the latter costs are included, the unit cost increases to 0.79 USD, which is still considerably less than the cost of alternative supply options, such as desalination.

Promoting regional cooperation

Water in international negotiations and treaties

The paper by Aaron T. Wolf, of Oregon State University in the United States, describes some results of his historical and statistical research on water-related treaties, which Wolf carried out using the International Crisis Behaviour data set. This research has led him to emphasize that solutions to international water conflicts must be creative and innovative.

Wolf points out that, in the initial stages of the Middle East peace process, everyone was thinking in traditional security terms, and they therefore focused on territory and resources. The stalemate over the red line in the Golan Heights, for example, was quite representative of the mind-set in the region. However, in Oslo II, territory was not tied to water as a negotiating factor. For example, land was returned to Jordan and the water on it was leased back to Israel (Israel farms the land and manages the water). The idea that one needs to keep territory to protect the water on it is no longer tenable.

Wolf's main point is that territory is no longer a driving force behind water negotiations; creative management is. For example, when negotiations over water between Israel and Jordan stalled over rights versus allocations, the mediators managed to break the deadlock by coining the term "rightful allocations," to be used in the treaty and now in international law. Although this was helpful in Israeli–Palestinian negotiations, it has not helped the Syrian question. Here, two opposing views of water resources under the Golan Heights (one Israeli and one Syrian) remain unresolved.

Wolf then outlines various options for resolving transboundary water issues, including

- Water management options to increase the supply and availability of water; and

- Adoption of a broader, regional perspective, from Egypt in the south to Turkey in the north (for example, Egypt could pipe water to the Sinai and Gaza, or the Litani water could be redistributed to Israel and the West Bank, and Turkish water imports could be relied on to improve regional water balances).

In evaluating these options, it is necessary to look at engineering, environmental, economical, and political variables. It is also important to take account of the "gut feeling" of all parties and where they stand vis-à-vis each other.

Wolf concludes by briefly reviewing technopolitical options, including

- The need to look at short- and long-term crisis management, to determine what are the areas most in need (for example, Gaza) and how outside parties can help now; and

- The importance of the planning time required for such big projects to become a reality.

A resolution

After the eight papers were presented at the conference, Husain Sadar, of Carleton University, presented a summary of the consensus reached at the workshop, listing specific conclusions and recommendations for future action. It was resolved by all that the first workshop should not be the last and only one. An Eastern Mediterranean Technical Water Advisory Group was seen as desirable and necessary for long-term, ongoing, cooperative effort. It was also seen as natural that Canada, given its technical expertise in water-related matters and technological resources, should pursue its role of organizer and facilitator. Al-Adwan then offered to host a second workshop at Yarmouk University in Irbid, Jordan, hopefully in the fall of 1999 or early in 2000.

Chapter 1

KEYNOTE ADDRESS: ACCESS TO WATER IN THE EASTERN MEDITERRANEAN

David B. Brooks

Introduction

One normally starts a talk by expressing one's pleasure at addressing the particular audience. In this case, I am not at all sure that I am pleased to be standing here. Typically, I talk about water in the Middle East before a Canadian–Jewish or Arab–Canadian audience, which, I can be 99% sure, knows a lot less about the subject than do I. Today I am standing before a group of individuals, each of whom is at least as knowledgeable in the subject as I am — a disconcerting prospect indeed!

I can only compare this experience to the time in Zurich, back in December 1992, when I was called on to deliver the keynote talk at the First Israeli–Palestinian International Academic Conference on Water. On that occasion, the organizers at least had the "courtesy" of neglecting to tell me that I was the keynote speaker until I arrived at the hotel in Zurich. The situation was no less disconcerting, but at least I had no time to agonize over the prospect.

Fortunately, there is an even better reason to look back now on the Zurich conference. That conference took place 9 months before the Israeli–Palestinian Peace Accord, and every word of its long and awkward title had been cautiously negotiated by a few courageous people. Yet, looking back, we can now see that the conference was seminal in demonstrating that water was an issue on which Palestinians and Israelis, researchers, policymakers, academics, and bureaucrats, could meet and reach equitable conclusions.

This workshop cannot, by its very nature, be seminal in the same sense. However, I do expect that it can and will be seminal in establishing a network, not merely for exchanging information, but for shaping ideas and promoting true science — what is coming to be called a "knowledge network." Certainly, both the timing and the geographic focus of this conference are right — and we have the two organizers, Professors Ozay Mehmet and Harvey Lithwick, to thank for that. Let me start by commenting on just why I think the timing and the geography are so important.

On timing

We have good reason to think that we are on the crest of a wave of renewed global recognition of the importance of fresh water to economic development, quality of life, ecosystem sustainability, and political stability.

A number of major reports have recently been released that focus on global or regional water issues. This attention given water has to be compared with that at the Rio Summit a few years back, or with that in the Agenda 21, which emerged from that Summit, both of which tried to focus on the Earth while all but ignoring water. I do not mean to imply that these reports are sanguine about freshwater supplies for an expanding world. Quite the contrary! However, they have established that water is (and probably always has been) the key natural resource constraint on development.

Conflict management studies increasingly identify lack of fresh water as a significant source of destabilization, migration, and local conflict, although I am glad to say that one hears less and less today of water as the cause of the next war in the Middle East. Water wars make for good headlines but not for very good economics or politics. Ever since Biblical times (as documented in Genesis), and no doubt long before, people have contended over water, and it is that contention, along with the various ways in which various interests have been managed, that will occupy a great deal of political economy over coming decades.

There is growing recognition that even if water is the very source of life, it is also an economic good. In particular, water pricing, as an issue of principle, is being put to rest. For example, in late 1998, the International Development Research Centre (IDRC), in collaboration with the International Water Resources Association, will be sponsoring a workshop on water and Islamic religious law

(Sharia), and pricing is one of the main points of discussion.[1] Of course, even as pricing becomes less of an issue of principle, pricing in practice needs to be studied that much more, including studies of rate structures, pricing by quality or delivery time, and appropriate ways to put prices on effluent water. At the same time, water can never be treated as just an economic good. Only the most callous free-market economists deny that everyone has the right to a minimum quantity of water regardless of their ability to pay. Moreover, as poets have long celebrated, water has an aesthetic dimension as well. We value water for itself, which makes it rather different from petroleum.

Finally water quality is being taken seriously — and just in time — in terms not only of pollution but also of direct loss. For many uses and in many places, degradation of water quality is equivalent to loss of water quantity. True, just as with energy, we can "cascade" water from higher to lower quality uses (as is happening now all over the Middle East, where used domestic water is treated and reused as irrigation water). However, the limits to such cascading are much greater with water than with energy, and, increasingly, we must give joint attention to water quantity and quality if we are not to find ourselves paying huge ecological and economic costs to clean up dirty water and find new sources of fresh water.

In short, we appear to be in a situation with water that bears some resemblance to the energy crises of 1973 and 1978. Happily, the analogy is more apt from an analytical than a political point of view. No country is in a position to cut off the supply of water, and no group of nations has oligopolistic power over it. However, every country knows that current practices for supply and delivery of water are outmoded. Consequently, wide agreement can be found for the position that new approaches to fresh water are needed, although there is very little agreement anywhere, intra- or internationally, on what those approaches should be.

On geography

The choice of the Eastern Mediterranean as the focus for our workshop is equally propitious.

---

[1] This was said in October 1998, before the event took place. I am pleased to report that it did happen as planned and that a book similar to this one, composed of the conference papers, will be published shortly.

Every one of those reports I referred to above identifies the Eastern Mediterranean as the area under the most immediate threat of big gaps between water supply and demand. If one takes the West Asian–North African region as a whole, average per capita renewable water supplies are around 1 250 m^3/year, which is just about one-sixth of the world average — this, in a region that is experiencing such rapid population and economic growth that per capita availability of water has fallen by half in just a quarter of a century! If anything, the elasticity of water use to income is increasing with time.

Worse yet, spatial and temporal distribution throughout the Eastern Mediterranean is dramatically uneven, which means that even the quantity of water available varies widely from place to place and month to month. One can plan for the more or less systematic geographic and temporal variations. But the really wide swings occur in the variation in rainfall from one year to the next, and these wider swings are much more difficult to account for in planning and investment. The Bible speaks of 7 years of plenty and 7 of drought, and, if one takes the number 7 as figurative, rather than literal, that is exactly the pattern we have in this region. Thus, the same area of the world where water supply is most limited is also the one with the highest peaks in uncertainty of supply.

The problems of the region are, by and large, shared problems, and it is entirely appropriate to look beyond the Israeli–Palestinian conflict or the Jordan Valley, which tend to dominate discussion to the neglect of countries with even more serious water problems, such as Cyprus and Jordan, and to the neglect of equally serious conflicts over shared international water resources, such as between Jordan and Syria. The Eastern Mediterranean is full of rivers and aquifers that flow across, along, or under international borders, and we may well want to spend as much time analyzing some of the smaller basins, such as the Orontes, as we have spent on the Nile and the Tigris-Euphrates.

At the same time, no other region in the world has so rich a tradition of concern for water. Nomads, pastoralists, and others who have traversed the area for centuries have an extraordinary understanding of surface water and how to use it efficiently and equitably. Their vocabulary includes more terms for water sources and courses than for any other topographic feature, terms that vary by rates and types of flow, elevation, accessibility, and permanence, among others. Equally impressive are the engineering works built to move water from the source to the fields or into the cities. Clearly, this is a heritage from which we still have much to learn.

In sum, the focus on the Eastern Mediterranean is as appropriate politically as it is analytically. As researchers and policymakers, we typically get to work first on the easier issues. In this case, we are perhaps reversing that procedure, and with good reason. If we can devise collaborative solutions for efficient, equitable, and environmentally sustainable water management in the Eastern Mediterranean, we can work them out anywhere on the globe.

Research directions

So much for the framework. I would say that this is a workshop of researchers, but I would not say "academics," as each of you keeps an eye on policy. Therefore, I want to take the remaining time to suggest topics to which researchers should direct their attention in order to have the greatest effect on policy, especially policy to promote collaboration on water issues. The list that follows is obviously idiosyncratic, but if it serves to stimulate debate, my aim will have been achieved.

To start provocatively, let me assert that the subjects simultaneously most relevant to policy and most seriously in need of study are neither technical nor narrowly economic, but socioeconomic, sociopolitical, and even sociopsychological. I do not mean to ignore the need for physical science or engineering. Clearly, we need better information on, for example, the role of vegetation in controlling runoff, aquifer hydrology, salt-tolerant crops, and barrage design for desert conditions. However, answers to those issues, vital though they are, can only serve as inputs to policy choices. Many projects for water demand management, water harvesting, and water reuse in arid and semi-arid areas have fallen far short of their potential because of failure to plan adequately for implementation and particularly for the socioeconomic, gender, and cultural dimensions that play so great a role in determining what projects can be implemented. The seemingly simple technique of water harvesting can serve as a metaphor for any activity where people and water interact.

To provide some structure, I will divide my suggestions for research into those for short-, medium-, and long-term issues. You will have to forgive me for introducing a few of my "hobbyhorses" along the way, which is to say, those topics on which I would particularly like to see some good research.

Short-term research needs

Immediate emphasis on research has to go in two directions, water demand management and marginal water supplies.

Water demand management

The key point to be made about water demand management is simple: every country in the region needs more of it, a lot more. None of the countries in the region is as efficient as is Singapore, for example, where a "total approach" has been adopted and where unaccounted for losses are now below 7%. The issue is not whether to manage demand but how:

- What measures make sense and where — in which sectors?

- What level of government should implement which measures?

- What techniques induce action by individuals, firms, and governments?

- What pricing will do, and what it will not do — whom will it affect adversely?

- Should prices reflect just the private value of water, or should social and environmental pricing concepts also be introduced — and, if the latter, how?

If we agree that every human being deserves some minimum quantity of water — say, 30 to 50 L/day — how can such water be provided to ensure that both delivery and use are efficient?

Hobbyhorse 1

As we undertake studies of water demand management, we must introduce the notion of an ecological demand for water. Even for those of us who see ecology as primordial, tough questions remain about just how big that demand is and the specific values it will satisfy. What values are lost, for example, when wetlands are drained by pumping, particularly wetlands in the semiarid parts of the Eastern Mediterranean?

Greater use of marginal water supplies

When speaking of "marginal water supplies," I mean to use the word *marginal* in two senses: first, in the sense that the individual increases in supply will be small and, second, in the sense that much of the water will be lower in quality. The questions here are similar to those for water demand management:

- How much water harvesting makes sense, and for whom — what are the blocks to greater use from rooftops, greenhouses, and fields — techniques that have been used in one form or another for centuries?

- What is the economic potential for using saline and other low-quality water in terms of the quantity and quality of crop output — to what extent is that potential limited by market or trade barriers?

- Should such approaches as rainwater harvesting be managed by individuals who can afford them, or should some incentives be introduced?

- What are the long-term effects of using marginal water on soils, crops, water bodies (including aquifers), and, of course, the human beings who grow the food and eat the crops?

Medium-term research needs

Once we get to research with a longer time frame, the questions begin to multiply, and the big ones are mainly institutional.

Water reallocation among sectors

Around three-quarters of all the fresh water in the Eastern Mediterranean is used for irrigation. It is now commonplace that over time much of that water will have to be transferred from agriculture to other uses. (In a few cases [notably Palestine], considerable potential remains for transferring water into agriculture.) However, just as with water demand management, simple recognition of a need raises more questions than it answers. Questions arrange themselves in three broad groups:

- The first group of questions focuses on how the transfers are to be effected: by market measures alone, some form of allocation, or a combination of these two. Are such reallocations meant to be for all time, or can they be effected on a temporary basis? Who will bear the costs of the transfers, and, conversely, what sorts of safety nets will be provided for those who decide to move (or are forced to move) out of farming?

- The second group of questions focuses on the farm systems themselves. A large number of people in the region depend on farming for their livelihood, and they need to know how to promote and protect the rainfed agricultural systems in the region. A lot of research remains to be done on appropriate farming systems for the Eastern Mediterranean. Part of the answer probably lies with supplemental irrigation, using shallow aquifers or water stored by rainwater harvesting. Supplemental irrigation can increase yields by a factor of as much as four in years with low rainfall, in comparison with nonirrigated fields, and it has the further advantage of permitting crop diversification, which is important to farm income.

- Finally, the third group contains all sorts of questions about direct and indirect trade. If a country is going to move out of agriculture, it is going to have to move into something else. In what products or services do Middle Eastern countries have a comparative advantage? Israel seems to have answered this question with high tech. Is that an answer for other nations in the region, too? If not, what else? Then there are questions about the sources of imports and security of supply. As Professor Tony Allen of the School of Oriental and African Studies at the University of London has emphasized, exporting oranges is equivalent to exporting water. He has also shown that as much water flows into the Middle East in food imports as flows down the Nile. Are there limits, political or economic, to the indirect trade in water? Should there be?

Hobbyhorse 2

Before we reallocate water too hurriedly, we must attend to a hot controversy. Some researchers, mainly at the International Water Management Institute (formerly, the International Irrigation Management Institute), insist that even if excessive water is used in irrigation, the surplus just returns to the soil, so that basin efficiency is much greater than field efficiency. The research question is not so much about whether the argument is right or wrong (to some extent it is certainly right) but about the extent to which it is right. We must learn more about the criteria for deciding whether intrabasin recycling is significant in any specific watershed.

Management of water at local levels

Almost all countries in the region treat water as a national resource, which is appropriate given the many roles that water plays as a public good; that is to say, water has a social value in excess of its private value. However, this says little about how centralized water management should be. We need to consider at least three broad kinds of question:

- First, given that at least some functions need to be centralized, how should the central agency be managed, and where should it be located in government? How can we give it the expertise and the independence it needs, yet make it ultimately subject to broad political direction? Israel learned that putting water under the Ministry of Agriculture was a bad choice, but is the new approach of putting it under the Ministry of Infrastructure any better? What experience is there with separating water policy from service and delivery of water — putting them into different institutions?

- Second, although some functions need to be centralized, we are learning that communities can do very nicely at managing local water systems, especially local irrigation systems. Very few experiments have been carried out in the Middle East, and we do not even know the criteria favouring a greater role for communities. Where do the "option boundaries" for local management lie? Are they different for surface and for groundwater? How do those boundaries vary with diverse systems of water rights and access? It is easy enough to argue for systems of common property in general, but how well will they work in more open economic systems with greater and more diverse demands on the limited water resource?

- Third, whatever the appropriate level of decision or service, how can a process of representation and participation be introduced into water management? Evidence is strong that, difficult though it may be, public review processes not only give stakeholders the "stake" that the term implies but also lead to more efficient and more equitable decisions. Democratic processes are far from the rule in the Eastern Mediterranean, and appropriate lessons could well be learned from countries such as India, where village-level management is the rule in some states.

Measures to deal with prolonged drought

Early in my talk, I mentioned that the wide variations in year-to-year rainfall are what make the design of appropriate water systems for the Eastern Mediterranean so tough. Some physical guidelines (as with the "red line" on the Sea of Galilee) have been designed, and it is accepted that, when supplies are really short, irrigation water is cut off before drinking water. However, analysis should be able to develop better methods for dealing with prolonged drought, such as offering water to certain consumers (most likely farmers) at discount prices, depending on their standing in a priority list for cut-offs in those years (for cutting off this use of water when supplies are below some specific point). California has had some success with "water banking." Does that approach have potential in the Eastern Mediterranean? Possibly some nonrenewable aquifers can be kept in reserve to be pumped in drought years and recharged in flood years.

Management of transboundary water

The most important water bodies and courses cross or lie along international borders. A lot of work is under way in this area, and knowing that some of you in this room are directly involved, I will restrict myself to just a single observation: if analytical and political issues are tough with transboundary surface water, they are even more so with aquifers. Ironically, a project (first presented to me at the First Israeli–Palestinian International Academic Conference on Water back in 1992) on joint management of the mountain aquifer may show the world how to do it.

Long-term research needs

I will restrict myself to a single proposal for the long term. Megaprojects are not very popular these days, and for good reason. Many have been expensive disasters, and almost all of them have been shown to produce fewer benefits than expected and to involve significant environmental and social costs not fully foreseen. Nevertheless, in the long term, and certainly by the second or third decade of the next century, one must at least consider options for a large supply of fresh water for this region of the world. There are not very many of them. Large-scale desalination, probably of brackish water to start, is the most obvious. Costs are high, but recent proposals for private-sector plants in Israel have used figures as low as 0.80 United States dollar/m^3 (if land is free and if by-product heat can be used). Such costs are not unreasonable for drinking water. The other big alternative involves international transfers, probably from Turkey, perhaps by pipeline or Medusa bags. What we need is some way to view the alternatives, along with the other region-specific megaprojects on the table (as with the Med-Dead and Red-Dead canals), to make each comparable with the others. This is not too difficult for the financial aspects, but it is much more so for the environmental and social ones, particularly when different value sets exist in different regions. (Just think how Canadians react to the prospect of exporting water to the United States!) In the same way, the analysis should show how such measures could take account of each nation's legitimate need for security of supply in the face of either hostile action or adverse climatic conditions.

Hobbyhorse 3

Two decades ago, when we were faced with an energy crisis that (at the time) seemed equally threatening, major advances in thinking were made using an approach called "soft energy paths." This approach turned conventional analysis upside down by emphasizing (1) energy demand, rather than supply; (2) energy quality, rather than quantity; and (3) backcasting, rather than forecasting. Based on my own research, I believe there is enormous room for an attempt to study comparable "soft water paths" — in effect, substituting "water" for "energy" in the previous sentence — and this is the subject to which I will devote such time as I can find for research over the next couple of years.

Water and the Israeli–Palestinian conflict

Finally, although I really believe in the geographic focus of this conference, what about the Israeli–Palestinian conflict? What can research on water do to assist in the resolution of this seemingly interminable conflict? Frankly, in my view, the answer is "not very much." Of course, research needs to be done in both Israel and Palestine, but apart from joint management of the mountain aquifer, no research I can think of is uniquely valuable to the resolution of the conflict. The problems are, and probably always have been, essentially political. None of the conflicts over water, not even the one between Israel and Syria over the Golan, is analytically difficult to resolve; indeed, compared with the other issues dividing these two nations, such as the rights of refugees, I believe water issues are quite manageable. There are even some win–win options left to pursue. Elisha Kally has shown that Jordan's lowest cost option by far for increasing water resources would be to store the high winter flows of the Yarmouk River in the Sea of Galilee and then to return them in the summer. The Technical and institutional procedures to permit this sort of banking should be almost straightforward for two countries now at peace with one another. In practice, mistrust and misunderstandings have blocked this option, although — as so often in Israeli–Jordanian affairs — more is going on behind closed doors than is admitted in public.

In summary, one could place the current Israeli–Palestinian situation in a good news – bad news framework. The good news is that Israeli–Palestinian water issues can be resolved by political will. The bad news is that they can only be resolved by political will.

So the final research question is what does one do to push reluctant, even recalcitrant, policymakers in the right direction. What is needed to make them see that we are all in this together, certainly in the long term, but even in the short term? I think we all have answers, but let me return to my opening remarks and suggest that conferences such as this one, with both researchers and policymakers together, are an important step. I am amazed at how much progress has been made since the 1992 Zurich conference and how much continues to be made despite the present difficulties. The Jordan–Israel Peace Treaty's annex on water is a case in point. One can question the specific allocations of water in the annex, but its design is almost a model for a water treaty. Equally encouraging are the efforts of the Water Resources Group of the Multilateral Middle East Peace Process, which, despite the current stalemate, continues to work to establish shared bibliographic and hydrological databases in the region. Perhaps most important for the long run, Israeli and Palestinian researchers who hardly knew one another 5 years ago are now working together, as on such partially IDRC-funded projects as joint management of the mountain aquifer and development of an environmental master plan for the Dead Sea.

I started with the statement that if collaborative solutions for water management can be worked out in the Eastern Mediterranean, they can be worked out anywhere on the globe. I will close with the statement that if collaborative solutions for water management can be worked out in Israel and Palestine, they can be worked out anywhere in the Eastern Mediterranean.

Chapter 2

ASSESSING LEBANON'S WATER BALANCE

Hussein A. Amery

Introduction

Lebanon is often thought of as having abundant water resources, but this view is rejected by virtually all of its public servants and politicians. General and widespread concern in Lebanon about water reflects the country's insecurity about perceived and real threats to its sovereignty over water resources, and this very attitude may affect the quality of water data collected, analyzed, and published. This chapter surveys existing water-related data, describes Lebanon's water balance, and outlines some of the reasons behind the deficiencies and discrepancies in the available hydrological data. One of this chapter's major findings is that there is a severe dearth of water data in Lebanon. This makes it difficult to manage water resources adequately in the present and to make meaningful projections into the future about potential use and availability of water. In the concluding section, a number of recommendations are made to help rectify the problem.

Challenges of data gathering

According to Bassam Jaber (Jaber 1997), Director-General of the Ministry of Electrical and Water Resources in Lebanon, the country installed 70 limnographic network stations in 1930. By 1974, one year before the civil war, only 20 of these stations were still in operation. Toward the end of the 19th century, the American University of Beirut had installed the first pluviometric station, and by 1974 a total of 150 such stations were functioning. After the end of the civil war, a mere 10% of these stations have been rehabilitated. Jaber added that Lebanon used to meter the water flow from all known springs in the country but ceased doing so in 1976. Snow cover is not metered, and the country's latest (now questioned) rainfall map was drawn by R.P. Plassard in 1970, almost three decades ago.

Lebanon's infrastructure for gathering water data is very weak to nonexistent, and this reflects negatively on the quality and accuracy of currently available data. Data used in this paper are drawn from unpublished government documents, interviews with senior public servants in the Ministry of Electrical and Water Resources in Lebanon, and official statements reported in the Lebanese media.

Hydrogeographical setting

Lebanon is a mountainous country with two parallel mountain ranges that run north to south, and between them lies the Bekaa Valley. These topographical features create an orographic effect. This results in heavy precipitation along the coastal plains and much less in the interior, and this rain shadow explains the notable difference in vegetative cover between the lush greenery of the coastal areas and the dry landscape of the interior. Lebanon's climate is generally Mediterranean, with abundant rainfall in the winter but dry summers.

The total area of Lebanon is $10\,422$ km^2 (about $4\,000$ square miles, or some $1\,040\,000$ ha). The country is made up of two principal hydrological regions: (1) the Mediterranean (or coastal) watershed, with an area of $5\,500$ km^2, which gives rise to 12 perennial rivers from the western slopes of the mountain ranges, flowing from east to west and emptying into the sea; and (2) the interior watershed, with an area of $4\,700$ km^2, which is the source of the Litani, Assi (Orontes), and Hasbani rivers. The country may be divided further into some 40 drainage basins of permanent or intermittent streams, whose flows depend on the topography of the watershed and the size of the mountain reservoirs that their sources feed (Fawaz 1967). Because the country has vast plains, generally good soil, and receives sufficient rainfall, more than one-third ($360\,000$ ha) of its land area is cultivable, of which about $190\,000$ ha is currently under cultivation.

The inland region of the Bekaa Valley has a continental climate. Its altitude is between 650 and $1\,000$ m above sea level, and this is the area from which the Litani and Assi rivers rise. Scarcely any human settlements are found above an elevation of $1\,600$ m, although many Lebanese mountains are higher than this. One peak reaches $3\,090$ m above sea level.

Rainfall is essentially a winter event. About 90% of all precipitation is received between November and April. January is the wettest month, and snow is frequently present in areas higher than 1 500 m above sea level. Precipitation varies spatially, as well as temporally. Precipitation in the highlands averages 1 500 mm/year, and the mountain peaks along the western ranges receive about 2 000 mm. Annual average precipitation in the northern Bekaa region, near Hirmil, is 250 mm; in Ba'albeck, 550 mm; and in Karoun and Marjoun, 700 mm. Yearly precipitation along the coast is 830 mm in the north, 800 mm around Beirut, and 700 mm in and around Sur (Tyre) in the south.

Water management in Lebanon

Fady G. Comair, President of the Board of the Litani Water Authority, presented a study toward the end of 1997 showing that the country has 2 200 Mm^3/year of surface water actually available and 2 600 Mm^3/year potentially available. Comair argued for the need to build 16 to 20 dams throughout the country and some mountain lakes to capture and use the state's water resources effectively. Comair's 1997 figures concur with Nasir Nasrallah's 1996 figures (An Nahar, 25 May 1996). However, other studies differ greatly from these "official" ones. Table 1 shows the most optimistic figures (Mallat 1982; United Nations 1992).

The figure of 2 600 Mm^3/year comprises 2 400 Mm^3/year of surface water and 200 Mm^3/year of groundwater. Nasir Nasrallah, Director-General of the Litani Water Authority, stated that the country can receive far more than 2 600 Mm^3 of water in a rainy year (one with above-average precipitation). This figure can also drop by as much as 50% if a drought lasts for a few years (An Nahar, 25 May 1996). Every 7 to 10 years, Lebanon does experience a drought, sometimes lasting for 3 or more years. (The 1988–91 drought, for example, reduced Lebanon's internally renewable water supply by 40%.) Naturally, droughts and lower surface-water flow result in higher levels of water pollution.

The United Nations (1992) report estimated that 2 557 Mm^3/year of available water are received during the wet season, and a small portion (818 Mm^3) is received during the dry season, for a total of 3 375 Mm^3 of available water. The volume of potentially usable groundwater is 600 Mm^3/year, but only 160 Mm^3 of this is used. Table 2 illustrates significant spatial and seasonal differences in distribution of water throughout the country.

Table 1. Hydrological budget of Lebanon: two perspectives (Mm³/year).

Factors	Water budget according to Litani Water Authority (Comair 1998a)	Water budget according to Mallet (1982)
Total precipitation	+8 600	+9 700
Evapotranspiration	−4 300	−5 075[a]
Percolation to groundwater and flow into the sea	−880	−600[b]
Flow into Israel		
Hasbani River	−160	−140
Groundwater flow to Huleh and northern Israel	−150	
Flow into Syria		
Assi River	−415	−415
Kabir River	−95	−95
Net available surface water	+2 600	+3 375

Source: Comair (1998a) and Mallet (1982).
[a]Includes groundwater seepage in Lebanon and from South Lebanon into Israel and into the sea.
[b]Excludes water flow into the sea.

Table 2. Geographic pattern of water distribution by season (Mm³/year).

Geographic area	Wet season	Dry season	Total
Western Slopes	1 958	515	2 473
Bekaa			
Assi Basin	54	43	97
Upper Litani basin	488	153	641
Hasbani River, springs, and other waters	68	96	164
Total	2 557	818	3 375

Source: Mallat (1982).

Mallat noted that a severe disparity occurs even between the seasons (Mallet 1982). In the dry season (Table 2), a mere 125 Mm³/year of water is available during the month of August, just when the need for irrigation is greatest.

Table 3. Water flow in the major hydrological zones of Lebanon in wet and dry seasons (Mm³/year).

Hydrological zones	Wet season	Dry season	Total
Water flow on the western slopes of Mount Lebanon			
Kabir to Beirut rivers	1 426	395	1 821
Damour to Zahrani rivers	489	102	591
Lower Litani (Khardali into the sea)	104	26	130
Ra's al Ain Spring (Sur)	15	13	28
Subtotals	2 034	536	2 570
Water flow in the Bekaa Valley			
Northern Bekaa (including Assi, excluding Litani)	248	264	512 [a]
Upper Litani (to Khardali)	488	153	641 [b]
Hasbani basin	112	40	152 [c]
Small springs	35	12	47
Subtotal	883	469	1 352
Total	2 917	1 005	3 922

Sources: Interviews with Lebanese officials, unpublished documents, and news reports.
[a] A total of 410 Mm³/year of the Assi flows into Syria.
[b] A total of 220 Mm³ are stored in the Karoun reservoir.
[c] A total of 140 Mm³/year flows into Israel.

Table 3 shows that the total amount of available water is 3 922 Mm³/year. However, the figure given by Mallat is 3 375, and the one provided by Lebanon's Litani Water Authority is significantly lower. Some of this statistical discrepancy may be due to the (unexplained) methodology, if any, followed by the various researchers. One does not know whether flows into neighbouring countries were counted as "Lebanese," nor whether groundwater within Lebanon or flowing across the border were included. Some, for example, included the 414.6 Mm³/year of the Assi River and the 95 Mm³/year of the Kabir River as water available to Lebanon. A total of 138 Mm³ of surface water flows from southern Lebanon into northern Israel and into the Huleh. However, although these figures are factually accurate, Lebanon is entitled to, or can exclusively use, only a small percentage of this water (See the section on the "Assi–Orontes River"). Finally, it is rare for authors to give the sources of their information on the availability or use of water flow.

The Litani River

Speaking at an international water conference at Kaslik University, Fady G. Comair (Comair 1998b, p. 2) said that "the cedar is the symbol of Lebanon, and the Litani represents life for the Lebanese." A prominent Lebanese hydrologist, Ibrahim Abd-el-Al, said in a lecture in 1950 that just as the Nile is the gift of Egypt, the "Litani is the gift of Lebanon" (Abd-el-Al n.d., p. 39). The Litani is often thought of as the key to Lebanon's future. These statements highlight the value of the river as perceived by the Lebanese people and illustrate the pivotal role that the Litani has in terms of the country's economic development, prosperity, and political survival. But this very perception has made it difficult to extract politically neutral, scientifically valid facts about the river. These observations are coloured by a perceived external "threat" to Lebanon's sovereignty over the Bekaa Valley in general and its control over the Litani River in particular.

The Litani River drops a total of 1 000 m from its springs to the Karoun Dam. The steepest descent is between Karoun and Khardali, where the river drops 600 m within a short distance. In its final stretch, the Litani flows rather gently and drops a total of 300 m over a distance of 50 km from Khardali to the sea north of Sur. The Karoun reservoir is used for irrigation, supply of potable water, hydroelectric generation, recreation (water sports, fishing, etc.), and tourism.

The civil war delayed water planning on the Litani River for a period of more than 20 years. Currently, the Council for Development and Reconstruction is studying the feasibility of supplying Beirut with water from the Litani River through the Jun tunnel (from the Karoun reservoir) before the water passes through the hydroelectric plant at Jun. This proposed water withdrawal will likely be 250 000 m^3/day in the first stage and 500 000 m^3/day by the final stage; the total annual volume will be 180 Mm^3 (Comair 1998a). This geographical reallocation of the Litani's water will contradict the spirit of Act 14522 (16 May 1970), which allocates the water of the Litani to the southern Bekaa (30 Mm^3) and the south (160 Mm^3) for domestic consumption and irrigation through canals at an elevation of 800 m above sea level.

Precipitation in the Litani's entire watershed averages around 700 mm/year. However, the lowest recorded precipitation in this area is 450 mm. Average precipitation in the upper Litani (above Karoun) is 800 mm/year within a watershed area of 1 600 km^2 (Comair 1998a). A total of 1 280 Mm^3/year of water is therefore received in this area, of which 60% seeps through to renew the groundwater or is lost to evapotranspiration. This leaves 500 Mm^3/year of surface water (in the upper Litani), of which 80 Mm^3/year is pumped out before reaching the Karoun, thus leaving 420 Mm^3 to reach the reservoir in an average year (Comair 1998a). This figure of 420 is based on a 35-year average (starting in the 1920s) but may decrease to 320 Mm^3 in years of drought, such as those that occurred in 1972–73 (Comair 1998a). Total annual flow in the Litani River is 700 Mm^3, given average precipitation levels (Comair 1998b).

Trying to trace the path of the Litani River during the months of July and August, this researcher was unable to find the spring south of the historic city of Ba'albeck, from which the river used to rise. The channel that had been carved by the spring's water was cracked; goats and sheep were grazing in it. This was probably a result of the massive pumping of groundwater that goes on in the Bekaa Valley. As the Litani River flows through the valley, the Mount Lebanon range rises in the river's watershed area, on the right, to an elevation of 2 620 m, and, on the left, to 2 400 m. The Litani watershed has an area of 2 168 km^2 (216 800 ha), 80% of which is located at higher than 800 metres above sea level. In this watershed, around 360 000 people live in 11 *caza* (administrative units) in which there are 161 settlements. These include four cities with populations of 10 000 or more, six towns with populations of 5 000 to 10 000, 57 villages with populations of 1 000 to 5 000, and the remaining settlements, with populations of less than 1 000.

According to Nasir Nasrallah (*An Nahar*, 25 May 1996), 100 Mm^3/year of water is diverted at a point near Markaba into the Awali River. Between Markaba and the town of Kilyah, 12 Mm^3/year is used for irrigation. Between Kilyah and Al Ghandourieh, 18 Mm^3/year is also drawn away for irrigation. Near the town of Kassmieh, 40 Mm^3/year is diverted for use in the Kassmieh irrigation project (Table 4).

Table 4. Water flow in the Litani River by hydrological zone (Mm³/year).

Hydrological zone	Winter	Summer	Total
Upper Litani (springs to Karoun)	360	130	490
Mid-Litani (Karoun to Khardali)	120	40	160
Lower Litani (Khardali to Kassmieh)	105	25	130
Total	585	195	780

Currently, according to Comair (1998a), the entire flow of the upper Litani is used for generating hydroelectric power. A total of 420 Mm³/year is used to generate 600 kW/h in the Litani River's three hydroelectric power plants: Markaba, Awali, and Jun. During the dry season, 30 Mm³/year of water is channeled from Markaba to help meet the needs of the Kassmieh irrigation project.

The Assi–Orontes River

The Assi River rises in an area north of the city of Ba'albeck and flows through Syria before entering Iskenderun (Alexandretta) and emptying into the Mediterranean Sea. The Al-Azraq spring is a very important Lebanese tributary to the Assi River; its annual flow is more than 400 Mm³ (see Table 1). In August of 1994, the Lebanese and Syrian governments reached a water-sharing agreement regarding the Assi, according to which Lebanon receives 80 Mm³ if the Assi River's flow inside Lebanon is 400 Mm³ or more during that given year. If this figure falls to lower than 400, Lebanon's share is adjusted downward, relative to the reduction in flow. Wells in the river's watershed that operated before the agreement are allowed to remain operational, but no new wells are permitted.

Water consumption

In 1966, the domestic and industrial sectors consumed 94 Mm³ of water, and the agricultural sector consumed 400 Mm³ (Table 5). By the mid-1990s, Lebanon was estimated to consume at least 890 Mm³/year of water, close to 50% of which was drawn from aquifers.

Table 5. Estimated water consumption and projected water demand in Lebanon (Mm^3/year). [a]

Year	Domestic	Industrial	Irrigation	Total
1966	94		400	494
1990 [b]	310	130	740	
1996	185–368	35–70	669–900	889–1 338
2000 [c]	280	400	1 600	2 280
2015	900	240	1 700	2 840

Sources: Jaber (1997); Comair (1998a); *Ad Deyar*, 6 Jul 1995, 5 Dec 1996.
[a] Water consumption data, with the possible exception of the 1966 data, are estimates. Thus, data vary a great deal depending on initial assumptions used.
[b] *An Nahar*, 25 Feb 1996.
[c] Nasir Nasrallah, as quoted in *An Nahar*, 25 May 1996.

Domestic consumption

Daily domestic water consumption was estimated at 165 L per capita in the mid-1990s. This figure is expected to reach 215 L by 2000 and 260 L by 2015 (Jaber 1997). Beirut currently uses 80 Mm^3/year of water, of which 30 Mm^3 comes from aquifers in the Damour region and 50 Mm^3 comes from those of Jeita. The capital is estimated to require 250 Mm^3/year of water, but its fresh and wastewater infrastructure is inadequate. This results in irregular supply of fresh water, especially during the summer and during outbreaks of waterborne illnesses when sewer lines break or leak. This has, at times, triggered civil disobedience and, consequently, tensions between the people and the police.

Industrial consumption

No data are available on the current or expected water needs of the industrial sector in Lebanon. In 1996, an estimated 71.4% of all industrial water used in the country was drawn from underground sources, and the remainder was drawn from surface sources.

Table 6. Potentially irrigable area within the Litani water basin.

Location	Area in hectares	Source of water
Ba'albeck	1 000 250	Ra's al Ain Spring Ain Hshbai
Zahle	2 100 900	Ain al Birdaouni Wadi Yafoufah
Shtura	400	Shtura Spring
Shtura-A'meek	5 000	Litani
Southern Bekaa	25 000	Karoun Reservoir (30 Mm^3) Groundwater (75 Mm^3) Shamsin Springs (33 Mm^3)
Rehabilitated lands in South Lebanon	33 000 (from a total of 75 300)	Karoun and Khardali reservoirs Springs in mid-Litani watershed
Jezeen (experimental plan)	930 (from a total of 1 500) 6 100	Karoun reservoir Karoun reservoir and springs south of Karoun
Other areas	(Unknown)	Small springs, some Litani tributaries, coastal freshwater springs, and groundwater
Total	74 680	Litani water basin

Source: Srour and Slaiman (1998).

Irrigation consumption

Of all the arable land in Lebanon, 146 000 ha was rain fed in 1996. The irrigated area was 23 000 ha in 1956 (10% of the then-cultivated area), 54 000 ha in 1966, 48 000 ha in the early 1970s, and 87 500 ha by 1993. According to studies conducted by the Food and Agriculture Organization of the Nations and by the United Nations Development Programme, the irrigated area of Lebanon is expected to rise to 170 000 ha by 2015 (Table 6), and this will require 1 700 Mm^3/year (Table 5). However, this projected demand could be kept down to only 1 300 Mm^3/year if water-saving approaches were implemented (Jaber 1997). Using traditional methods, 10 000 m^3 or more of water is required to irrigate 1 ha of land in a given year. More efficient methods would cut this required volume to 6 000 m^3.

The potential increase in irrigable land in the southern Bekaa Valley is from 23 000 to 25 000 ha (Table 6). Around 20% of this land, especially lying along river banks, requires drainage. In 1972, 10 000 ha in the southern Bekaa Valley was irrigated. Another 13 000 ha is scheduled for irrigation. The water required to irrigate these two areas is 140 Mm^3/year, of which 30 Mm^3/year will be drawn from Karoun Lake, 74 Mm^3/year from groundwater, and 36 Mm^3/year from other surface-water sources (United Nations 1992). To rehabilitate and irrigate lands in the southern Bekaa Valley (between Bir Illyas and Jib Janeen), the national Litani Water Authority straightened and deepened the river's channel between 1970 and 1972. This project was later delayed as a result of the civil war. During the summer of 1998, the Authority completed building an irrigation canal that had been dug in the mid-1970s without ever going into operation. Also during the summer of 1998, the Authority started to extend the existing canal system beyond its then northern reach.

There are plans to irrigate 6 000 and 4 000 ha, respectively, in the Hermel and Akkar regions of the northern Bekaa. A total of 33 000 ha is slated for irrigation in southern Lebanon. This includes 1 200 ha near Saida and the currently irrigated area of 6 000 ha in the Kassmieh region. In the coastal plain, 58 000 ha can be irrigated by coastal rivers and aquifers.

Administrative and legal structures

One of the troublesome and urgent hydrological issues in Lebanon pertains to groundwater management. Order in Council 144 (10 June 1925) states that public property is any that may, by its very nature, be used by many people or for the benefit of the general public. Regardless of how much time may have passed in ownership or use of a certain land resource, such properties may not be sold or profited from, and they include surface and groundwater, lakes, rivers, and lake and river banks. However, in the late 1960s, the legislation was amended to exclude wells drilled on private lands with an output of less than 100 m^3/sec. Such wells must not pump water that possibly belongs to someone else or feeds into a river. Innumerable wells are found throughout Lebanon, especially in the Litani watershed, and for various reasons enforcement of existing laws is very lax to nonexistent. Some wells were dug during the civil war to meet the water needs of nearby rural communities. Until the government water infrastructure is rebuilt, it would be difficult to imagine the state enforcing the law on such wells. In addition, the average age of the poorly paid staff at the Ministry of Electrical and Water Resources is 55, and this figure is increasing yearly because of the current hiring freeze. The Ministry also suffers from a shortage of technical and managerial skills.

Water quality

The quality of potable water is of growing concern among citizens, popular organizations, and the nongovernmental organizations that operate in the country, as is the quality of near-shore seawater. One of the areas in the public eye is the industrial zone in the Litani watershed near the town of Zahle (Table 7). The majority of plants (with five workers or more) located in the Litani watershed (except for cattle and poultry farms) are grouped in seven clusters, most in the vicinity of Zahle. This area, according to a 1996 survey of industries (N'khaal et al. 1998), contains a total of 36 factories, including 15 plants for processing food and beverages and 6 plants for manufacturing nonmetallic products (glass, ceramics, etc.). The Lebanese Ministry of the Environment is monitoring the industrial waste of about 100 factories, with the choice of factories based on the size and type of activity and operation engaged in. Of these, four are located in the Litani River watershed, producing food, paper, and nonmetallic products.

Conclusions and recommendations

Lebanon has been making ambitious and energetic efforts to rebuild its war-ravaged economy and has accomplished significant advancements in this direction over the last several years. However, the dearth of up-to-date and reliable hydrological data makes it very difficult to plan adequate and sustainable management of the country's water resources into the 21st century. This data deficiency is critical, because economic-development planning must consider the current hydrological picture to make sensible projections about growth and water needs. Reliance on 1930s data and 1950s infrastructure and instruments for gathering water data is an obstacle to the sustainable management of Lebanon's water resources.

Table 7. Water quality along the Upper Litani River: EPA maximum upper limits and results of Litani water sample from southeast Zahle.[a]

Substance measured	EPA maximum limits	Southeast Zahle readings
ph		8.21
BOD		71
COD		62
TSS		18
TDS		207
Ta		190
Tn		1.4
TP		2
OIL		<1
Phenol		0.02
CN-		0.024
THM		0.47
As	5	0.45
Ba	100	<0.05
Cd	1	
Cr	5	<0.01
Co		<0.01
Pb	5	0.15
Hg	0.2	600
Ni		<0.01

Source: N'khaal et al. 1998.
Note: BOD, biochemical oxygen demand; COD, chemical oxygen demand; THM, trihalomethanes; TDS, total dissolved solids; TSS, total suspended solids.
[a] Units are in milligrams per litre, except for ph level.

One would be hard pressed to identify any administrative district in Lebanon in which all villages and neighbourhoods receive a continuous water supply, have indoor plumbing facilities, and are connected to a sewer system. Nasrallah (quoted in *An Nahar*, 25 May 1996, p. 9) stated that "there is not a single village

Table 8. *Caza* (Administrative Districts) and sewer-system networks in Bekaa Province.

Caza	Towns (n)	Towns with SSNs (n)	Share of towns with SSNs (%)	Share of people with SSNs (%)
Ba'albeck	94	2	2	11
Zahle	49	12	24	52
Western Bekaa	40	1	3	8

Source: N'khaal et al. (1998).
Note: SSN, sewer-system network.

or city in Lebanon that receives an uninterrupted residential supply of water, nor is it possible to [meet the desire of residents to] irrigate all agricultural lands." The Director-General of the Litani Water Authority was essentially saying that Lebanon's water resources are barely sufficient to meet current, let alone future, needs (Table 8). This is one of the key arguments used by Lebanese officials for rejecting proposals to divert water to other countries. It must, however, be noted that water production and distribution systems are inadequate to meet the rapidly rising demand for the water resources that Lebanon does possess. Population growth and internal displacements (a legacy of the civil war) have put excessive stress on a water infrastructure that was designed to pump and deliver daily only 50 to 100 L per capita (Jaber 1997), well below the levels of current and future demand.

The following is a list of some of the actions needed to bring Lebanon's water management capabilities into the 21st century:

- Refurbish, enlarge the capacity of, and expand the geographic coverage of, existing infrastructure. (This should be done with the aim of improving water delivery and reducing water loss, which will also protect groundwater from wastewater contamination.)

- Upgrade and update the skills and other capacities of staff.

- Strengthen management in order to develop more efficient use of limited financial resources. (Stronger management is likely to be an efficient way to collect water subscription fees.)

- Update hydrological maps at a scale of $1 : 2\,500\,000$, showing major catchment areas, drainage lines, and patterns, rivers, lakes, and other major and minor water bodies, as well as rainfall distribution patterns.

- Produce hydrogeological maps at a scale of $1 : 2\,500\,000$, showing groundwater flow patterns, water-quality levels, aquifer boundaries, existing development, and areas for potential future development. (Drawing these maps can be done with the help of very useful relevant data obtained from meteorological satellites, Landsat Multisprectral Scanners, and other such devices. There is, however, no substitute for land surveys, which provide, among other things, human field confirmation of technologically gathered data.)

- Provide drainage to areas at high risk of soil salinization.

Lebanon's water management team has a long, slippery road to chart before it can develop the human, financial, and technological resources desperately needed to protect and make full use of the country's water resources.

References

Abd-el-Al, I. n.d. Al Majmouah al Kamelah li A'maal Ibrahim Abd-el-Al [the complete works of Ibrahim Abd-el-Al]. 3 vols. The Friends of Abd-el-Al, Beirut, Lebanon. [In Arabic and French]

Al Hajjar, Z.K. 1997. Al Meyah al Lubnaneyah wa Assalam fi Asharq al Awsat. Dar al A'ilm Lil Malayeen, Beirut, Lebanon.

Comair, F.G. 1998a. Sources and uses of water from the Litani Basin and Karoun Lake. Paper presented at the Workshop on Pollution in the Litani Basin and Lake Karoun, and Environmental Problems in the Western Bekaa and Rashaya, 9–10 May 1998.

——— 1998b. Litani water management — prospects for the future. Speech given at the International Conference on International Water Law and Water Education, 19–20 Jun 1998, Kaslik University, Jounieh, Lebanon.

Fawaz, M. 1967. International Conference on Water for Peace, Beirut, Lebanon. Vol. I. Government Press, Washington, DC, USA. pp. 293–299.

Jaber, B. 1997. Water in Lebanon: problems and solutions. Public lecture given in the Department of Hydrology, Purdue University, Lafeyette, IN, USA, Apr 1997.

Mallat, H. 1982. Meyah Lubnan Naft Lubnan [water in Lebanon–petroleum in Lebanon]. No. 6. Department of Law, Politics and Management, Lebanese University, Beirut, Lebanon. [In Arabic]

N'khaal, Saad al Deen; Ali, N.; Smaha, E. 1998. The role of wastewater in polluting the Litani water and Lake Karoun. Paper presented at the Workshop on Pollution in the Litani River and Lake Karoun, and Environmental Problems in the Western Bekaa and Rashaya, 9–10 May 1998.

Srour, S.; Slaiman, B. 1998. Water management plan for the Litani River and Karoun reservoir. Paper presented at the Workshop on Pollution in the Litani River and Lake Karoun, and Environmental Problems in the Western Bekaa and Rashaya, 9–10 May 1998.

United Nations. 1992. Ground water in the Eastern Mediterranean and Western Asia. United Nations Natural Resources, New York, NY, USA. Water Series No. 9. ST/ESA/112.

Chapter 3

Evaluating Water Balances in Israel

Harvey Lithwick

Introduction

Israel provides an interesting case study of how the understanding of a nation's water balances can change with advances in technology, growing economic sophistication, and evolution in internal and regional politics. In most circumstances, water balances have been viewed as exogenously determined — the difference between available sources and uses, both of which were deemed to be largely mechanistically predetermined. Over the past decade especially, research in Israel has revealed that the issues are much more malleable, particularly with regard to the role of market forces. As a result, what was once viewed as an impending crisis has now been more realistically addressed as essentially an allocation problem, one that is not simple, but much less apocalyptic. It has been learned that the potential for dealing with a variety of regional conflicts over water can be significantly enhanced with the wise application of management and pricing regimes. Indeed, there has been a radical revision in domestic policy with respect to water within Israel over this period, and it is to be hoped that this same change in thinking will help contribute to alleviating long-standing disagreements at the regional and international levels.

NB: This paper was previously published as a working paper by the Negev Center for Regional Development, Ben-Gurion University of the Negev, Beer Sheva, Israel. I is reproduced here with the permission of the author. The author gratefully acknowledges the research assistance of Ms Tilly Shames, a visiting graduate student from the Norman Paterson School of International Affairs at Carleton University in Ottawa, Canada, and Mr Dovi Wilensky, from the University of Santa Cruz. This chapter was also made possible through the assistance of a number of individuals and groups who kindly provided information, references, advice, and cautions. The author alone is responsible for their use here. I would specifically like to thank the following: Tony Allen, School

Traditional factors shaping water balances

The traditional approach to water resources was to focus on the quantitative "stock" of water, with particular attention paid to additions to, and removals from, that stock. Removals from stock were shaped by allocation mechanisms, which in most countries reflect the interplay of powerful interests. I begin with a summary of this sort of water accounting in Israel. Then I provide a brief overview of the historical background and conclude with a discussion of several currently salient issues.

The entry point: supply of water

Israel has three major storage basins for its stock of water. One is rainwater and melting snow, primarily from Mt Hermon, which enter the upper Jordan River and then flow into the Sea of Galilee. The other two are the coastal and mountain (Yarkon) aquifers (Figure 1). These three sources account for almost two-thirds of Israel's current annual water supply of just less than 2 000 Mm³/year. The rest is made up equally from smaller aquifers, especially in the Western Galilee and the Arava–Negev region, and from recycled and brackish water (Table 1).

These sources are primarily dependent on annual replenishment through rainfall. This entry point is problematic because of various factors, the most important being short-term climatic variability and the possibility of longer term periods of significant declines resulting from prolonged drought. The Sea of Galilee has had annual inflows ranging from a low of 100 Mm³ in drought years (most recently in 1991) to a high of 1 500 Mm³ (Kliot 1994). These phenomena impose

of Oriental and African Studies, University of London; Shaul Arlosoroff and Hillel Shuval, Hebrew University, Jerusalem; Raphael Bar-El, Faculty of Management, Ben-Gurion University; Zvi Eckstein, Tel Aviv University; Franklin M. Fisher, Massachusetts Institute of Technology; Nava Haruvy, Agricultural Research Association, The Volcani Center, Bet Dagan, Israel; The Israel Information Center; Moshe Israeli, Israel Water Commissioner; Mrs Esti Landau, The Armand Hammer Fund for Economic Cooperation in the Middle East, Tel Aviv University; Malaika Martin and Barry Rubin, *Middle East Review of International Affairs*, Begin-Sadat Center; Thomas Naff, University of Pennsylvania, Middle East Water Information Network; Uri Regev and Moshe Justman, Department of Economics, Ben-Gurion University of the Negev, Beer Sheva; Uri Shamir, Josef Hagin, and Ms Ella Offenberger, Water Research Institute, University of Haifa; Boaz Wachtel, Consultant, Tel Aviv; Amos Zemel, A. Issar, Gideon Oron, and Hendrik Bruins, The Desert Research Institute, Sde Boqer.

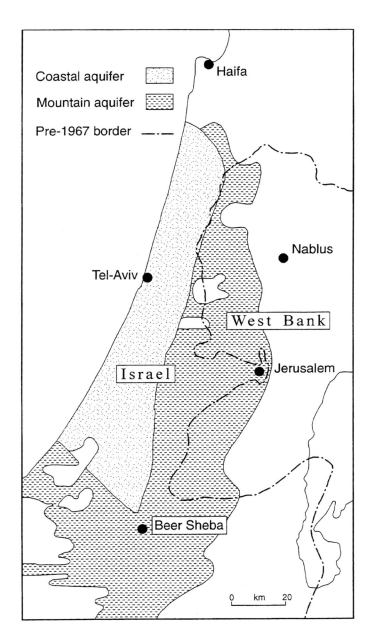

Figure 1. Israel's principal water-supply sources. Source: GOI (1998).

Table 1. Sources of water supply in Israel, 1996.

	Mm³/year	% of total
Major sites		
Sea of Galilee	512.4	27
Coastal aquifer	418.0	22
Yarkon aquifer	326.1	17
Subtotal	1 256.5	65
Other sites		
Negev–Arava	89.0	5
Western Galilee	85.6	4
Other	162.2	8
Subtotal	336.8	17
Low quality		
Dan sewage water	140.9	7
Brackish water	130.7	7
Other	61.3	3
Subtotal	332.9	17

Source: GOI (1998).

on planners the need to make appropriate risk allowances when estimating future requirements. On the other hand, only part of the inflow manages to find its way into the water supply. Evaporation from the Sea of Galilee amounts to more than one-third of its annual inflow (Table 2). Also, losses resulting from leaky pipes, especially in urban areas, have been estimated at about 5% of the total annual production. In some cities, it has been estimated that up to 50% of the supply may be lost because of such leakage (Kliot 1994).

In Israel, there are long-standing stocks of water in the Fossil Desert aquifers (under the Negev and Sinai), which, at present, provide some 30 Mm³/year but are estimated to be able to provide several hundreds of cubic metres of water

Table 2. Key storage basins for groundwater, 1995.

Major source	Mm³/year	Comments
Sea of Galilee		
Effective stock	600	
Inflows:		
Jordan River	494	For years 1980–85
Runoff	216	
Precipitation	65	
Other	37	
Subtotal	812	
Outflows		
Evaporation	294	
Downstream	42	
Into water supply	500	
Subtotal	–24	Resulting in pollution and salinity
Coastal plain aquifer		
Effective stock	320	
Net flow	–96	Resulting in pollution and salinity
Mountain aquifer		
Inflow		
Precipitation	350	
Negev aquifer		
Outflow	30	

Source: GCI (1998).

a year (Issar 1998[1]). However, these stocks are not rechargeable, and this means that the draws on them are essentially nonreversible, which may partly explain why this source has not really been exploited.

---

[1] Issar, A.S. 1998. Global change and water resources in the Middle East: past, present and future. Unpublished manuscript.

The National Water Carrier, one of Israel's most important infrastructure projects, moves a very large proportion of Israel's water supply from the north of the country to users in the northern Negev (Figure 2).

Finally, there have been increased efforts to reuse water, that is, to put used water back into stock. This entails a lower level of water quality, which affects the allocation process, a subject that I shall return to below, as treated, recycled sewage water has been the major "new" source of water in Israel. Total sewage water produced in Israel amounted to 453 Mm^3/year in 1990. At that time, just more than one-third was treated for use in irrigation, but plans are to increase the volume of treated water for irrigation to 292 Mm^3 by 2000 (Kliot 1994).

Eckstein et al. (1994) provided more comprehensive estimates of Israel's potential water supply by source:

- Underground reservoirs, 1 250 Mm^3/year;

- Jordan and Sea of Galilee, 640 Mm^3/year;

- Lower Jordan–Yarmuk, 85 Mm^3/year;

- Streams and springs, 130 Mm^3/year;

- Treated wastewater, 460 Mm^3/year;

- Total, 2 570 Mm^3/year.

Most of these water sources are under dispute with Jordan, the Palestinian Authority, and Syria (Fisher 1995). The rough estimates of the annual water flow under dispute are as follows: Jordan, 600 Mm^3/year; Yarmuk, 500 Mm^3/year (250 of which flows south of Syria); the mountain aquifer, 600 Mm^3/year; and total, 1 700 Mm^3/year.

Traditional practice has been to search for new water sources to deal with a perceived shortage, and a number of schemes have been advanced over the years. There remains very active debate about capital costs, operating costs, and,

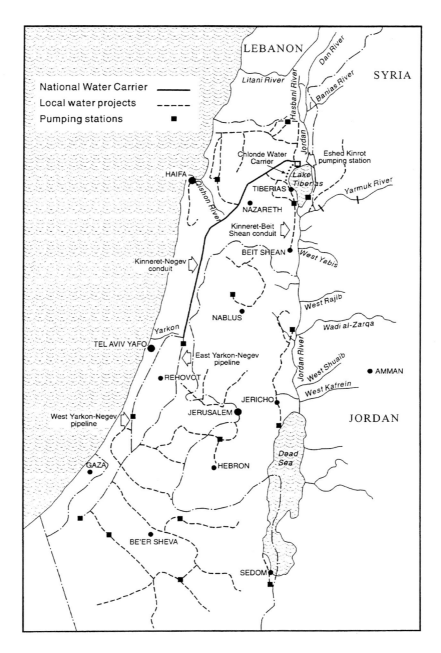

Figure 2. The National Water Carrier and related water products. Source: Kliot (1994)

of course, security of supply, when considering possible projects. Only a brief review of these schemes is possible here. The following is a summary of their key characteristics and, where appropriate, estimated costs per cubic metre.

- More intensive use of brackish waters (already being implemented);

- More intensive capturing of rainwater (a potential 160 Mm^3/year), including use of microdams (Laronne 1996);

- Desalination of seawater (with cost estimated at 0.75–1.00 United States dollars (USD)/m^3 in 1992 prices);

- Importation of water from the Litani River in Lebanon (geopolitical constraints);

- Importation of water by sea from the new Manavgat depot in southern Turkey (estimated costs have exceeded the minimum 0.75 USD/m^3 for desalination; James Cran (Cran 1994), a proponent of the Medusa-bag technique [using a ship to tow chains of huge plastic bags filled with water] estimated the cost of this solution at 0.18 USD/m^3, but this is far below the price that the Turkish authorities wish to charge [see Nachmani 1995]);

- Overland importation of water from Turkey (see Wachtel n.d.[2]) via Syria and the Peace Canal (this scheme is not costed, and it has major geopolitical constraints);

- Importation, by canal, of Nile water to Gaza and the Negev (a cost of 0.40 USD/m^3, but with geopolitical constraints); and

- Canals to link the Mediterranean or the Red Sea to the Dead Sea (Figure 3); the estimated costs, excluding delivery, range from 1.00 to 2.00 USD/m^3 (Bar-El 1995).

---

[2] Wachtel, B. n.d. The Peace Canal Plan. Mimeo.

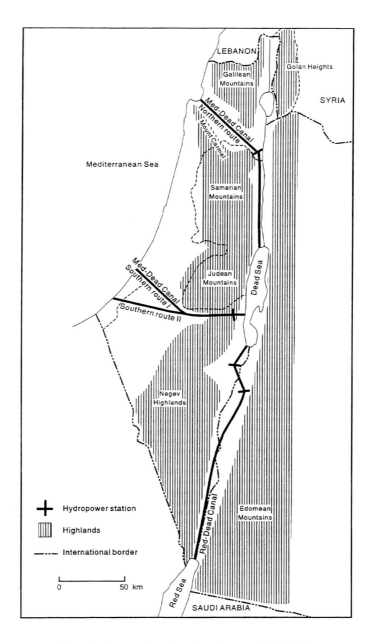

Figure 3. Proposed canal projects. Source: Hillel (1994).

Although widely used as the basis for choosing among alternatives, such cost comparisons do not even constitute cost-effectiveness evaluations. At best, they estimate direct costs, with little attention to accounting explicitly for external benefits or costs, and they would appear to use widely varying discount rates, etc. To the best of my knowledge, no systematic social cost–benefit analysis (SCBA), the most appropriate project analysis tool for such comparisons, has been undertaken.

It should be stressed that the availability of alternative water supplies at different costs makes the aggregate supply curve of water a rising-step function, rather than a vertical one, as is commonly claimed. The highest relevant cost is generally believed to be that of desalination — it will likely dominate all other major proposed sources in the next few decades. There is some dispute as to when it will become a cost-effective option. The Harvard team, headed by Fisher, concluded that desalination, compared with currently available options, would not likely become economically feasible until 2020. Shadow prices of other fresh water on the Mediterranean coast, where such plants would be located, are not expected to rise to more than 0.70 USD/m^3, in 1990 prices, until 2020. The cost per cubic metre resulting from the canal projects is higher, and, because of this, the canal option is dominated by the coastal desalination alternative (Fisher 1995).

The use stage: demand for water

The dominant user of water in Israel is the agricultural sector. Despite the decline of this sector in relation to the national economy, down from 11% to 5% of gross national product since the founding of the state, and despite the virtual elimination of Israeli agricultural products for export, down from 60% to 4%, agriculture has grown significantly in absolute terms, with important implications for overall water use. The area under cultivation has almost tripled from 162 000 ha to 445 500 ha, and the amount of farmland under irrigation has increased nine times, from 28 350 ha to 255 000 ha. As a counterbalance, new irrigation techniques have lowered water use per hectare by one-third. Nevertheless, agriculture still accounts for about 64% of all water consumed (MEWIN 1998). Of this, kibbutzim consume 44%, and moshavim (an organizational form involving cooperative management but private ownership) use 33% (Lindholm 1995). The strong political organization of these entities plays an obvious and important role in influencing the mode and levels of water allocation. By contrast, the share of domestic and urban use stands at about 30%, and industrial use is at 6% (GOI 1998).

The different prices charged to users have reflected a bias toward subsidizing water-intensive agriculture. At present, the price continues to differ by up to a factor of two. This influences not only the allocation of water among users but the overall rate of the use of water as well. The most recent average prices we have found are as follows (for October 1996, since raised):

	ILS	1996 (USD)	1992 (Eckstein et al. 1994) (Actual USD)
Agriculture	0.62	0.19	0.17
Industry	0.83	0.26	0.11
Domestic	1.12	0.35	0.50 to 1.22
Wastewater	0.50	0.16	

Note: ILS, Israeli new shekel. The exchange rate in 1996 was 3.2 ILS = 1 USD (in 1999, 4.0903 Israeli new shekels [ILS] = 1 United States dollar [USD]).

For efficient, realistic pricing, users should pay the marginal social cost of water delivered to their particular location; this marginal direct cost averages about 0.35 USD/m^3. As it is now, there is a major subsidization of agricultural and industrial water use by taxpayers, but the greatest burden is on domestic-sector users, who in 1990 provided an overall subsidy of some 200–250 million USD for water use (Kliot 1994). A major reason for these cross and overall subsidies is that in Israel a politically responsive state monopoly controls the allocation of water and investment in water projects, which ensures inefficient allocation of water supplies. I will elaborate on this issue in the following section.

Recent estimates project the annual growth in demand for water in Israel at about 30 Mm^3/year, mostly because of urban and industrial expansion. However, official projections, particularly those of Water Planning for Israel Ltd (TAHAL), have had to be subjected to some upward revisions because of changes that had to be made to their underlying assumptions. The most systematic of the revised estimates, until recently, were those of Eckstein et al. (1994).

For the household sector, the figures take into account the accelerated growth in population resulting from the wave of immigration from the former Soviet Union, which added some 700 000 to previous population growth estimates.

It also meant a higher than expected growth rate for the Palestinian population. The end result was an increase in household consumption by 39 Mm^3 in 1990, which will be 52 Mm^3 by 2010.

For manufacturing, most demand is concentrated on food processing, quarrying, and the chemical industry, much of which is located in the south of the country. However, not much growth in demand is expected for the industrial use of water, although the West Bank and Gaza are showing modest industrial growth, especially in the food production sector.

For agriculture, the estimates of use should be based on price assumptions. TAHAL stuck to volume estimates, albeit while implicitly reflecting an acceptance of higher prices, and projected a decline in quotas for agriculture that amounted to between 17 and 25% in total for all water and 55% for fresh water. This projection would be offset to some extent by an increase in agricultural consumption of water in the West Bank.

Naturally, there is serious concern over the net balance between inflows and outflows, discussed above, because over time the continued net withdrawals (or deficits) will deplete the water stock or render it less usable, as a result of qualitative deterioration. Kliot (1994) estimated these accumulated net deficits for up to 1990 (Table 3). These net flows should be seen in the larger context of the existing stock to provide some perspective on the nature and extent of the problem. One such attempt to estimate the relationship between stocks and flows at a key site — the Sea of Galilee — is summarized in Appendix 1 of this chapter, which shows that net annual flows constitute between 12% and 14% of the total stock of water in the lake. This is not meant to imply that all of this stock is available for extraction at times of severe shortage, because depletion below some red line will cause severe environmental damage to the lake and lakeshore. In recent years, the level has receded very close to that red line, and there is, therefore, legitimate concern over any annual deficit.

The politics of the water allocation process

With the whole stock of water in mind, decisions must be made regarding the allocation of these supplies among various sectors (agricultural, industrial, and residential); and the location of these various users is another factor. Allocations always reflect political considerations, together with economic realities. Allocation of water based on economic considerations tends to promote efficiency in both the

Table 3. Stacks and flows of water from major sources, 1990 (Mm^3).

Source	Net outflow	Overuse	Accumulated deficit
Underground			
Coastal aquifer	240–455	34–80 (1980–90)	100–1 400
Local aquifers	23–280		Small
Mountain aquifer	300–330	50 (1980–90)	300–350
Surface			
Sea of Galilee	575–950	25 (1980–85)	140
Floods and treated sewage	200–230		
Total	1 890–2 311		1 570
Water losses	60–100		
Balance	1 790		

Source: Kliot (1994).

production and consumption of water, as well as increasing the efficacy of major new water project investments, but other modes of allocation do not. In Israel, economic considerations long played a secondary role, thereby exacerbating the scarcity problem. However, over the past decade, Israel has made some substantial progress toward taking water allocation away from agriculture and putting it toward other uses that yield higher returns.

There is an interesting semantic phenomenon in referring to the use of water by various sectors as "demand." Economists define demand as the amount that consumers would like to purchase at alternative prices, but most of the forecasts for water demand appear to be based on quantitative extrapolations of water volume, ignoring or at best underestimating the importance of pricing and income effects on that demand. The consequence is that if prices charged are substantially below their true competitive equilibrium, the estimated volume demanded, and hence used, will be much higher than it would be economically efficient and socially optimal to supply.

Water is, for most purposes, what economists call an "intermediate input." As such, the value of water other than for household use is based, not on the utility derived from direct consumption of the water itself, but on the value of the goods and services it helps to produce. If the outputs are valued in competitive markets, the value of the water can be readily estimated. Where they are not, such as in highly protected agricultural markets, the value of water is more difficult to measure and must be derived through shadow pricing. It has been estimated that the value of the marginal product of 1 m^3 of water in agriculture is between 0.15 and 0.30 USD (Arlosoroff 1997). Economic rationale would therefore allocate water to such a use if its delivered costs were less than its value. As the delivered cost is a function of location, the net effect would be to reduce water use for agriculture in remote regions. Similarly, it would tend to reduce the production of those crops whose value, per unit of water used, is relatively low. Clearly, this would affect a wide variety of agricultural interests.

Although reduced consumption is therefore an appropriate goal, all too often it is promoted by the public sector in advocating specific technologies. Appropriate pricing is the preferred alternative, because it would encourage the most cost-effective technologies to be introduced at the appropriate time within the various sectors. However, the dominant users of water in the agricultural sector, represented by the Association of Farmers, have resisted such a policy orientation for perhaps obvious reasons.

To the extent that the allocation process is based on noneconomic considerations, it is very likely that use will bear a limited relationship to overall community valuations and real resource costs. That is not to argue that political considerations are not important — security of food and energy supplies for a security-conscious state like Israel is indeed of great importance. However, it may well be that misallocation of water actually contributes to less security by wasting a relatively scarce resource and making peaceful solutions to interregional water disputes more difficult. Recent attempts to impose a more rigorous cost and price discipline should go a long way to encouraging more efficient use of water (Arlosoroff 1997). Over the long run, efficient pricing also ensures that investments in the supply and use water are also efficient.

Calculations of water's scarcity value in Israel

In Israel, reallocation of water use is achieved in the face of the long-standing interests described above, mostly as a result of the accumulation of evidence on the costs and benefits linked to water use. It is useful to begin with extraction costs. Quantities extracted and costs of extraction (in 1992 prices) from other sources (common pool) are as follows (Eckstein et al. 1994):

	Volume (Mm³)	Price (USD/m³)	Conflict
Southern coastal aquifer	49	0.42	Gaza
Yarkon aquifer, north	90	0.14	
Yarkon aquifer, south	110	0.20	
Gilboa	131	0.31	
S'dom, Dead Sea	84	0.12	
Ramallah	25	0.57	West Bank

As for the aggregate supply prices of water, the marginal costs of extracting water have been estimated by Bental (1996), as presented in Table 4.

Based on Eckstein's estimate of the cost of water from the mountain aquifer (0.50 USD/m³), some important orders of magnitude of the benefits to be derived have been clarified. If this represents the efficient price of water, then the value of the estimated 2 000 Mm³/year used is about 1 billion USD, or 1.7% of the gross domestic product (GDP) of the entire region. For a highly efficient water-use regime to emerge, the allocation would have to change dramatically. Water in the northern Negev (and Gaza) costs about twice as much as in Galilee.

Table 4. Marginal water-extraction costs.

Volume (Mm³)	Marginal cost (USD/m³)	
	1991 ILS	USD
0	0.34	0.15
700	0.46	0.20
1 100	0.68	0.30
1 400	0.91	0.40
1 700	1.25	0.54
1 900	1.60	0.70
2 000	1.82	0.79

Source: Bental (1996).
Note: ILS, Israeli new shekel. The 1991 exchange rate was 2.3 ILS = 1 USD (in 1999, 4.0903 ILS = 1 United States dollar [USD]).

The efficient use of water would require a 10% cut in allocation in the north and a 40% cut in that in the south, mostly to agriculture and primarily for marginal crops that have a very low value added per unit of water input. The study estimates that if water had been priced based on efficient allocation, total water consumption would have fallen by 296 Mm^3 in 1992, to 1 779 Mm^3, that is, by 16%. The price of water would have risen by 0.30 USD/m^3, and the quantity used in agriculture would have fallen by 10–15%. The price of water in the south would have risen 170%, and the quantity used in agriculture would have fallen by 25–30%. Based on this evidence, current efforts to move major amounts of agriculture to the Negev appear to be extremely ill considered.

An important study by Gideon Fishelson (Fishelson 1993), at Tel-Aviv University in 1993, provided the first set of elasticity estimates for household water demand. He estimated the long-term income elasticity at between 0.2 and 0.4. The price elasticity was estimated at between −0.05 and −0.15. Based on these fairly low elasticities, the author argued that even at very high prices, household consumption would be very unlikely to decline below the benchmark current consumption of 110 m^3/year.

Historical trends and recent estimates

Long-term trends in water balances since the late 1950s reveal that the agricultural demands grew steadily until 1983 but then declined dramatically — by almost one-third — between 1983 and 1990. But this trend was sharply reversed during the first half of the current decade. It is the domestic sector that has undergone steady long-term growth, offsetting whatever savings were realized in agriculture over the previous 10 years (Figure 4). We should note, however, that on a per capita basis, overall water consumption has declined substantially in Israel (Figure 5), no doubt in large part because of the slowdown in agriculture's consumption of water since the mid-1980s.

Table 5 presents the most recent estimates of water balances in Israel, projected to 2040. They make a major improvement in water allocation planning possible because they are based on more realistic projections of demand, supply, and the use of efficiency-based allocation procedures. Overall, these procedures have granted Israel a period of perhaps a decade in which to find more fundamental solutions to its long-term water requirements. Knowing that these solutions will take

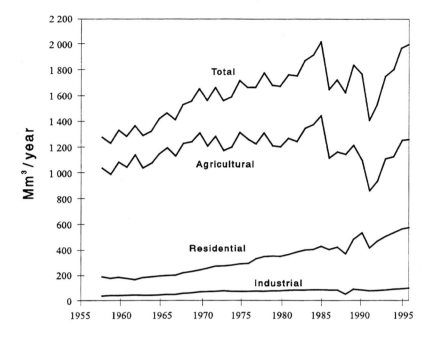

Figure 4. Historical trends in water use (Mm³/year). Source: GOI (1998).

a number of years, it is a matter of some urgency to begin the planning in the very near future. Ideally, a combination of approaches should be considered in order to avoid undue reliance on any one technology. For example, a 10-year contract to purchase water from Turkey, coupled with the development of pilot desalination plants on the Mediterranean, could be considered, but only after appropriate SCBAs had been conducted.

Water: a heterogeneous product

I have, to this point, assumed that water is a homogeneous product, but what complicates the story of water use is that it can and does exist at different levels of quality. Some of its uses do not require the highest level. Clearly, a system that optimizes water use will attempt to allocate such quality-differentiated supplies in

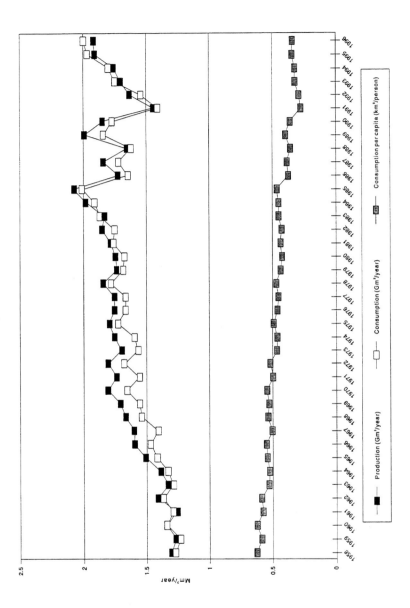

Figure 5. Historical trends in production and consumption of water in Israel. Source: GOI (1998).

Table 5. Israel's current and projected water balances (Mm^3).

		1990	2000	2010	2020	2040
Annual inflows						
Israel and West Bank	Groundwater	1 060	1 090	1 100	1 100	1 100
	Jordan Basin	660	670	670	670	670
	Floodwater	40	50	70	80	70
	Losses	−40	−40	−30	−25	−25
	Subtotal	720	1 770	1 810	1 825	1 815
	Reused water	198	296	418	651	1 071
	Total	1 918	2 066	2 228	2 476	2 886
Annual outflows						
Israel	Municipal	481	654	774	915	1 151
	Industrial	106	130	155	183	255
	Irrigation	1 200	1 200	1 200	1 370	1 920
	Subtotal	1 787	1 984	2 129	2 468	3 326
West Bank	Municipal	36	71	133	204	379
	Irrigation	100	155	190	280	300
	Subtotal	136	226	323	484	679
Gaza (net use)		43	43	69	94	147
	Total	1 966	2 253	2 521	3 046	4 152
Net flows Israel and the West Bank		−48	−187	−293	−570	−1 266

Source: Israel Water Study for the World Bank (cited in GOI 1997).

the least-cost manner, a process that is already under way in Israel, but, for perhaps understandable reasons, it encounters significant resistance. For example, because of the need and desire to supply an extremely high standard of drinking water to people, all water delivered to households must meet this standard, even though the bulk of it is not used for drinking or cooking, but for bathing, cleaning, laundering, and even watering the geraniums. Water for direct human consumption constitutes a minuscule portion of total household water use. Methods to encourage alternative modes of delivering drinking water could conceivably reduce significantly the need for high-quality water for other household uses and hence lower its costs. Also, a practice of encouraging direct household recycling of "gray" water for garden use would promote important efficiencies.

The allocation system implicit in the Telem study for TAHAL in 1988 (TAHAL 1988) was based on similar considerations. The plan was to reduce freshwater consumption from 1 800 Mm^3/year to 1 600 Mm^3/year. Household demand for fresh water would rise by 480 Mm^3, and water going to agriculture would fall by 660 Mm^3. Recycled water would be allocated in much larger amounts to the agricultural sector, both as a substitute for the lost high-quality water and to enable further expansion. The current distribution of water by level of quality is provided in Table 6, which indicates that the targets have been achieved.

The roles of technology and economics

A key question is what is the value of water to the Israeli economy. Using the price of desalination as the maximum willingness to pay and the shadow (efficient allocation) price of 0.50 USD/m^3, the net value of the common pool available is estimated at 200 million USD/year, or less than half of 1% of Israel's GDP (Eckstein et al. 1994). The net rents from the common pool are slightly less than 100 million USD, which could serve as the basis for financing water projects.

The scale of desalination to date is modest. Most of the plants are in the remote Eilat area, and they meet more than half of that city's needs. As we have seen, in other parts of the country, the process is not cost-effective, nor does it appear likely to be in the near future. A major factor contributing to the high cost of desalination is its heavy energy requirements, the costs of which tends to be understated and the security implications of which tend to be ignored.

Efforts to enhance rainfall through seeding clouds with silver-oxide crystals have been made over the Sea of Galilee for the past two decades. The result has been an increase in annual rainfall in that area by almost 20%.

Existing water supplies can be augmented through the use of new technologies, as Israel has demonstrated in numerous fields. On the one hand, improvements in drilling techniques have made once inaccessible stocks an important component of annual supply. On the other hand, microsprinklers and, more recently, drip irrigation with computerized control systems have made much more

Table 6. Water use by user and water-quality level, (1995) (Mm^3).

Use	Fresh water	Effluents	Brackish	Total	Mekorot share	Mekorot (%)
Agriculture	898	227	86	211	747	62
Domestic	578		2	580	444	77
Industrial	111		25	136	94	69

Source: Arlosoroff (1997).

efficient use of existing water supplies in agriculture. About 20% of consumption for irrigation has been reduced by these methods. New technologies for using brackish water in the agricultural sector, without diminishing yields, have had beneficial impacts as well.

A major new source of water is treated household and industrial effluents. More than 100 m^3/year from this source is now being used in agriculture (cotton and fruit growing), but another 200 m^3 is still discharged into groundwater or into the sea, owing to the absence of storage facilities.

A decade-old program involves building artificial lakes (120 to date) to collect surplus winter runoff. The water in these lakes can be used, not only for irrigation, but also for recharging aquifers; and the lakes can be used for storing water in transit between uses and locations. How these innovations came about — the result of responses to scarcity, signaled partly at least through rising prices — remains to be fully analyzed. Certainly, the subsidization of many uses of water has retarded such innovative processes, but it is expected that recent reforms will give much freer rein to imaginative solutions.

An alternative means of augmenting water supplies is through importation, rather than production, of especially water-intensive, low value-added food supplies. With the opening of global food markets and intense competition among suppliers, countries in the region, such as Egypt, have been able to forestall a potential water crisis by importing food. For decades, Israel followed the opposite path, subsidizing via the price of water agricultural production and exports, effectively encouraging the export of the water that it took to grow the food (Allan 1998a). By shifting to food imports, by more carefully allocating water supplies (especially to high-cost locations), and by avoiding crops with low value-added water input, the overall social impact of an increase in the price of water toward its true scarcity value can be substantially mitigated.

The other side of this coin is less comforting. The use of fertilizers and insecticides, which in part permits agriculture to make do with less water, also contributes to the reduction of water quality. Kliot reported that according to the Water Commissioner, most of the water for household use is below the official quality standard, especially with regard to its high nitrate content, which does not conform to internationally accepted standards (Kliot 1994). The most severe effects of overpumping, as a response to shortages, are seen in Gaza, where the level of contamination of the groundwater is extremely high (Kliot 1994).

Furthermore, Mekorot Water Company's distribution system is very energy intensive. Energy represents more than one-quarter of the company's operating costs, and the company uses 8% of the power generated by the Israel Electric Corporation. One suspects that the associated environmental costs (air pollution) directly attributable to water provision are not yet being factored into its price. Offsetting this is the fact that with the increased use of treated effluent water for agriculture, fewer pollutants enter urban streams and the sea, reducing the already alarming levels of environmental damage (with its high social costs) in densely populated areas. Groundwater is affected in the rural areas, where such water supplies are used, but lower densities of population in rural areas will tend to reduce the net social cost of this pollution transfer.

The role of geopolitics

The core problem facing Israel is that its major source of surface water (the Jordan River) and its underground water sources (the two aquifer systems) are also claimed by other jurisdictions. The Jordan River has a complex system of sources and distribution, as can be seen in the schematic presented in Appendix 2 of this chapter. The two major actors are Israel and Jordan. The Palestinians are involved primarily through their claims on the aquifers adjacent to their territory, adding a second dimension to the debate.

These intercountry conflicts can be broken down into two distinct issues: (1) the issue of who owns the water, or what is legally known as property rights; and (2) the issue of spillovers or externalities, situations in which one party's actions have implications (positive or negative) for another. An efficient allocation of water does not depend on property rights, so long as the water is properly priced and the rights to it are freely traded. There is no such simple solution for externalities, because parties acting in their own self-interest tend to avoid taking these effects into account. If the external effects are costs to others, the result is that too much is produced (the classic example being road congestion). If the effects are benefits, too little is supplied. A third-party or common management system is required in such circumstances to ensure that efficient amounts are produced and exchanged.

For a common management system to be completely effective, it must involve all actors with an interest in the system and the ability to affect it. Cooperation must take place in the form of joint action plans, commissions, and treaties, based on a regional approach to watershed planning that involves all riparian states and regional actors with an interest in the water source. In the case of Israel and its neighbours, this requires basin-wide cooperation, involving Jordan, Lebanon, the Palestinian Authority, and Syria, together with Israel. The peace treaty of 1994 between Israel and Jordan provided for a division of water resources without the involvement of the other riparian states. The supply of water affected by this agreement could be diminished and joint cooperative plans under way could be derailed by other parties with access to, and an interest in, this water source.

Is Israel able and willing to approach the management, distribution, and allocation of these shared water resources as a shared task? Any sharing of water will be seen as reducing Israel's ability to meet its own water needs, even if doing otherwise would entail infringing on others' right to meet their needs. Israel is in a unique position of having a great deal of control over the distribution of both underground and surface sources, and this affects its neighbours. However, Israel remains highly vulnerable to potential actions by the other riparian states as well. Israel is heavily dependent on two contested supplies: the 430 Mm^3/year that it receives from the mountain aquifer and an additional 305 Mm^3/year of fresh renewable water from the Golan, totaling 735 Mm^3/year of Israel's 1 587 Mm^3/year total freshwater consumption (see Table 6). The mountain aquifer poses a big challenge. The Palestinians are unable to expand their own water resources in this region. Extensive groundwater development in the West Bank would threaten coastal wells because of increased saltwater intrusion from the sea (Wolf 1995). Moreover, any pollution of this underground source of water will result in a net loss of water available for Israel's population. Therefore, to protect its scarce sources of water, Israel believes it needs to control groundwater exploitation and prevent contamination.

Despite the many innovations noted above, it is far from certain that the long-term water needs of the region will be met as demand continues to expand. To date, in the absence of frameworks for cooperative action, innovations have been made based on narrow, inward-looking criteria. For example, Jordan constructed its East Ghor Main Canal system, which runs along the east coast of the Jordan River, to serve agricultural needs in its country while Israel developed its National Water Carrier system, starting at the Sea of Galilee and carrying water throughout the country. These and other initiatives began to interact, resulting in growing tension. The war of 1967 is a key example of escalated tension leading to conflict. The inclusion of water issues in the multilateral Israeli–Palestinian peace negotiations highlights the importance of this issue to the future development of this region and the resolution of conflict.

The potential for cooperation is certainly there. In addition to nonconventional water resources that can be developed unilaterally, there is scope for joint research and innovation programs. Moreover, short-term water needs can be alleviated through interbasin transfers of water. Options include diverting water from the Litani River to the Sea of Galilee (providing 100 Mm^3/year to Israel, Jordan, and the West Bank), channeling water from the Nile to the Jordan watershed (resulting in 500 Mm^3/year), sending water from Turkey to the Jordan watershed by pipeline (1 100 Mm^3/year), and using Medusa bags to ship water from Turkey (500 Mm^3/year) (See Appendix 3 of this chapter). Longer term cooperation could focus on regional initiatives, such as desalination projects.

The degree to which these projects are possible will depend on the willingness of these states to cooperate for the sake of enhancing water resources to meet the water needs of the region as a whole. The combination of a need to expand water sources and a dependence on shared water sources should provide a powerful incentive to cooperate. A peaceful resolution to conflict in the region would increase the chances of successful implementation of any and all proposed projects. At the same time, pursuing these initiatives may encourage further dialogue and cooperation among riparian actors. As such, cooperation over water may contribute to, and benefit from, an environment of peace.

Water as a symbol

Perhaps the greatest barrier to finding reasonable solutions to the so-called Middle East water crisis, at both the national and regional levels, is the symbolism attached to the resource. In Israel, it is intimately bound up with the early Zionist views about land and the importance of agriculture in settling and claiming it:

> Water for us is life itself. It is food for the people, and not food alone. Without large-scale irrigation — we shall not be a people rooted in theland, secure in its existence and stable in its character.
> — Prime Minister Moshe Sharret, in 1952 (quoted in Feitelson and Haddad 1994, p. 73)

These views persist to this day in the subsidization of water for agriculture, which transfers costs to other users, as well as to the economy as a whole, in terms of wasted resources. The approach of focusing on water volume alone has led many to conclude that current rates of overuse are plunging the region into a crisis. Such a view has been justifiably ridiculed by no less an authority than a former Israeli water commissioner, Dan Zaslavsky, who pointed out that "there are local and temporary shortages because it's not the highest priority of the countries involved; that's all!" (quoted by Nachmani 1995; see also Zaslavsky 1997). The traditional view is changing, and more rational allocations, using more appropriate prices and more realistic water-quality mixes, are emerging on the part of the water authorities themselves.

One adjustment mechanism has been stressed by Allan (1998a), namely, importing "virtual water" at low cost in the form of food products from region's that have a comparative water advantage. Another is the major reduction in water use in Israel, from 2 000 Mm^3/year in the mid-1980s to less than 1 600 in less than a decade, primarily through an increase in productivity in agriculture, occasioned by higher prices, which reflect growing scarcity. Unfortunately, the update on that story is a bit less optimistic, as the last few years have seen a sharper increase than was anticipated, with total consumption in 1996 once again approaching 2 000 Mm^3/year (see Figure 4).

On a regional basis, issues of sovereignty enter in, and water has been at the centre of long-standing, major disputes. Once again, too much focus has been on water volumes alone and allocating them among the various states with conflicting claims. But these huge claims are based on existing patterns of allocation that fail to allocate water in terms of its scarcity value (shadow price), as strongly expressed by Nachmani (1995). In other words, few dare to question the demands or needs being claimed, but they are certainly exaggerated because use is priced below true scarcity value. Allan goes so far as to claim that in the Middle East, "water almost everywhere is treated as a free good" (Allan 1995, p. 344). Moreover, as Fisher (1995) and others have shown, the implicit value of the water in conflict is surprisingly small and appropriate solutions are feasible. The value of this water is estimated by Fisher at no more than 110 million USD, which will rise to some 500 million USD (in 1990 prices) by 2010 (Fisher 1995).

Of course, this argument ignores the possibility that water may not be the cause but the symptom of more basic conflicts, so that managerial–economic solutions are beside the point. Nevertheless, a less symbolic approach to water has helped Israel achieve substantial efficiency gains in its national water use, and a similar approach applied regionally may offer some hope for collective action at that level.

Prospects

Israel has obtained a modest window of opportunity to deal with its own and the region's water needs. By moving toward a policy of efficient allocation, it has been able to restrain the growth in demand, even with a very rapid surge in population resulting from immigration from the former Soviet Union in the early 1990s. The immediate challenge for Israel is to further reduce the share of fresh water going to the agricultural sector. The old mode of administrative allocations will not do the job, as it is subject to historical interests incapable of readily accepting the burden of such a change. One alternative would be to extend the current initiative to divert fresh water from irrigation and replace it with treated effluent, but this option is limited by quantitative and qualitative constraints and could only serve as a partial solution. Fortunately, market mechanisms have been proposed, including tradable rights and the use of appropriate scarcity pricing. If adopted, these changes would have a profound and beneficial impact on the whole water economy. Adoption of similar policies by neighbouring countries could provide temporary relief for the region as a whole.

Two critical steps are required if the region is to avoid serious impending difficulties. One of these would be to find the means to operate regionally (that is, multinationally), whatever the political circumstances may be, to deal effectively with the externalities intrinsic to this scarce resource. The other would be to make more effective use of the price mechanism, a move required to ensure efficiency in managing the stock of water. The advantage of this would be that it tends to be less political and less bureaucratic and can therefore help avoid the problems that are bound to occur in any multinational effort at regional cooperation (Eckstein et al. 1994; Fisher 1995).

Even with efficient pricing and regional cooperation in management, the growth in demand, early in the next century, will once again bring serious water shortages to the fore. A number schemes to add to Israel's and the region's water supply are being vigorously promoted by their respective proponents: desalination, a variety of canal schemes, importation of water from Turkey, and capture of runoff, to name a few. Despite substantial analysis of each proposal in isolation, I have been unable to discover a serious attempt to rigorously compare the full set of social costs and benefits from these alternatives, a question amenable to the tools of SCBA. Water projects have been the first, and still the most important, field for the successful application of this methodology (El-Bihbety and Lithwick 1998). Water authorities would be well advised to underwrite some baseline studies in this area to enable Israel to identify and implement realistic solutions.

Whatever schemes are adopted, progress toward regional cooperation in meeting short-term requirements can provide important institutional mechanisms for positive-sum long-term solutions as well. Acting collectively as water buyers, we can keep import prices down. Acting collectively as project developers, we can capture economies of scale and positive externalities.

Paradoxically, Israel's recent successes in dealing with its short-term challenges may lead it to resist those region-wide collaborative efforts that could do much to alleviate the longer term problems. Viewed constructively, a move toward regional cooperation may, in the short run, not only provide opportunities for low-cost, long-term solutions but also play a useful role in creating a less hostile geopolitical environment for everyone.

Appendix 1. Recent water balance of the Sea of Galilee (Mm³/year).

Source of flow	Inflow	Plus	Minus	Outflow
Flow into Sea of Galilee	544			
Rainfall over sea		65	−270	−474
Flow from local runoff		70		
Springs in and around sea		65		
Evaporation from sea surface				
Outflow to lower Jordan River				
Total volume of sea	4 000			

Source: Murakami (1995).

Appendix 2. Stocks and flows of water from major sources (Mm³).

Source	Net outflow	Overuse	Accumulated deficit
Underground			
Coastal aquifer	240–455	34–80 (1980–90)	100–1 400
Local aquifers	23–280	Small	
Mountain aquifer	300–330	50 (1980–90)	300–350
Surface			
Sea of Galilee	575–950	25 (1980–85)	140
Floods and treated sewage	200–230		
Total	1 890–2 311		1 570
Water losses	60–100		
Balance	1 790		

Source: Kliot (1994).

Appendix 3. Comparison of alternative water import schemes.

Mode	Volume (Mm³/year)	Price (USD/m³)
Litani to Israel	100	0.14
Nile to Israel	500	0.20
Turkey, overland	1 100	NA
Turkey, Medusa bag	500	0.21

Source: Wolf (1995).
Note: NA, not available.

References

Allan, J.A. 1995. Striking the "right price" for water? Achieving harmony between basic human needs, available resources and commercial viability. *In* Allan, J.A.; Mallat, C., ed., Water in the Middle East: legal, political and commercial implications. I.B. Tauris Publishers, London, UK. pp. 325–346.

———— 1998a. Middle East water: local and global issues.
http://endjinn.soas.ac.uk/Geography/Waterissues/Papers/9508ta_01.html

———— 1998b. Watersheds and problemsheds: explaining the absence of armed conflict over water in the Middle East. Middle East Review of International Affairs, 2(1).
http://www.biu.ac.il/soc/besa/meria/journal/1998/issue1/jvol2no1in.html

Arlosoroff, S. 1997. The Public Commission on the Water Sector Reform. The Truman Institute of the Hebrew University, Jerusalem, Israel.
http://atar.mscc.huji.ac.il/~truman.sarlopap1.htm

Awerbuch, L. 1988. Desalination technology: an overview. *In* Starr, J.R.; Stoll, D.C., ed., The politics of scarcity: water in the Middle East. Center for Strategic and International Studies, Scranton, PA, USA. Westview Special Studies on the Middle East. Ch. 4, pp. 53–64.

Bar-El, R. 1995. The long term water balance east and west of the Jordan River. Ministry of Economy and Planning, National and Economic Planning Authority, Kiryat Ben-Gurion, Jerusalem.

Bental, B. 1996. A dynamic macroeconomic model of water uses in Israel. The Economic Quarterly, 43(1) (1996), 7–20. [In Hebrew]

Cran, J. 1994. Medusa bag projects for the ocean transport of fresh water in the Mediterranean and Middle East. Paper presented at the 8th World Water Congress, 21 Nov, Cairo, Egypt.

Eckstein, Z.; Zakai, D.; Nachtom, Y.; Fishelson, G. 1994. The allocation of water resources between Israel, the West Bank and Gaza: an economic viewpoint. The Pinchas Sapir Center for Development, Tel Aviv University; The Armand Hammer Fund for Economic Cooperation in the Middle East, Tel Aviv, Israel.

El-Bihbety, H.; Lithwick, H. 1998. Cost–benefit analysis of water management mega projects in India and China. *In* Bruins, H.J.; Lithwick, H., ed., The Arid Frontier. Kluwer, Dordrecht, Netherlands. pp. 295–317.

Feitelson, E.; Haddad, M., ed. 1994. Joint management of shared aquifers. Harry S. Truman Research Institute, Hebrew University; Palestinian Consultancy Group, Jerusalem, Israel.

Fishelson, G. 1993. The Israeli household sector demand for water. Tel Aviv University; The Armand Hammer Fund for Economic Cooperation in the Middle East, Tel Aviv, Israel.

Fisher, F.M. 1995. The economics of water dispute resolution, project evaluation and management: an application to the Middle East. Water Resources Development, 11(4), 377–390.

GOI (Government of Israel). 1997. Partnership in development 1998, Nov 1997, Chapter 2. Presented at the Middle East / North Africa Economic Conference, Doha, Qatar. Government document, Jerusalem, Israel.

———— 1998. Water in Israel, consumption and production, 1996. Ministry of National Infrastructure, Water Commissioner, Tel Aviv, Israel. [In Hebrew]

Hillel, D. 1994. Rivers of Eden: the struggle for water and the quest for peace in the Middle East. Oxford University Press, Oxford, UK.

Kliot, N. 1994. Water resources and conflict in the Middle East. Routledge, London, UK.

Laronne, J., ed. 1996. Reservoirs as sources of water for the Negev. Report of a conference, 15 Feb 1996. Ben-Gurion University of the Negev, Beer Sheva, Israel; Jewish National Fund, Jerusalem, Israel; Ministry of Agriculture and Rural Development, Jerusalem, Israel; National Water Company (Mekorot), Tel Aviv, Israel; Water Authority, Jerusalem, Israel. [In Hebrew]

Lindholm, H. 1995. Water and the Arab–Israeli conflict. In Ohlsson, L., ed., Hydropolitics: conflicts over water as a development constraint. Zed Books, London, UK.

MEWIN (Middle East Water Information Network). 1998. Israeli water use and exports. http://gurukul.ucc.american.edu/ted/; cited 13 Sep 1998.

Murakami, M.; Musiake, K. 1994. The Jordan River and the Litani. In Biswas, A.K., ed., International waters of the Middle East: from Euphrates to Tigris to Nile. UNEP,United Nations Environment Programme, Paris, France, Ch. 5, pp. 117–155.

Nachmani, A. 1995. Water jitters in the Middle East. Studies in Conflict and Terrorism, 17(1), 67–93.

TAHAL (Water Planning for Israel Ltd). 1988. A Master Plan for the Israeli water market. [In Hebrew]

Wolf, A.T. 1995. Hydropolitics along the Jordan River: scarce water and its impact on the Arab–Israeli conflict. United Nations University Press, New York, NY, USA.

Zaslavsky, D. 1997. Solar energy without a collector for electricity and water in the 21st century. Lecture to the Austrian Academy of Sciences, Vienna, Austria, May 1997.

Chapter 4

WATER BALANCES IN PALESTINE: NUMBERS AND POLITICAL CULTURE IN THE MIDDLE EAST

Samer Alatout

Introduction

This chapter focuses on four issues related to Palestine's water balance and regional hydropolitics. The second section discusses various attempts at estimating
Palestine's water balance, each claiming scientific legitimacy and technical authority.[1] The conclusion of this section is that estimates of Palestine's water balance
are both technically uncertain and reflective of the cultural and political contexts
within which they are produced. The issue of local hydropolitics — using water
resources to manage local political conflict — is not dealt with in this chapter, but
this is an important topic and will be the focus of another paper.

Because of the complexity of water balances, the third section turns to a
discussion of the political grounding of water estimates, an issue seldom considered in literature on water balances in the Middle East. Using new insights
from the field of science and technology studies (STS), it is argued that the seemingly neutral and objective language used to discuss Palestine's water balance is,

NB: I would like to thank the Norman Paterson School of International Affairs at Carleton
University, especially Ozay Mehmet, for inviting me to participate in this workshop, and the
International Development Research Centre of Canada, especially David B. Brooks, for funding
my participation. I am thankful to the workshop participants and to Saul Halfon for their
comments on an earlier draft of this paper. Finally, I am particularly grateful to David B.
Brooks for his insightful comments on it.

[1] This section synthesizes many of the studies published in the 1990s on Palestine's water
balance and water politics. These studies can be found among the works cited in the
bibliography.

more often than not, imbued with politics.[2] This is not the same as the claim often heard that water experts are captured by political interests. In all but a few rare cases, it is doubtful that this conspiratorial image holds true under rigorous investigation. General insights from STS, together with particular insights gleaned while researching the culture of water expertise in the Middle East, demonstrate that knowledge about water is necessarily produced within, and shaped by, institutional, cultural, and political contexts. Bearing this in mind, the difficult task of those looking for a workable water-sharing regime between Israel and Palestine is to unpack and make explicit the politics of technical jargon, rather than adding additional layers of obscurity.

The fact that most water resources in Palestine are shared with other parties makes them subject to international law, which is the third issue to be discussed in this study. Although it is agreed that the international legal principle of historic right is appropriate for the Palestinian–Israeli context, this principle is insufficient on its own. Another international legal principle, that of the equitable use of shared water resources, is at least as important as historic right in determining Palestine's water share. Neither of these two principles should be excluded from any new attempt to design a water-sharing regime. Putting these two international legal principles into operation simultaneously is going to require constant negotiation and collaboration among Israeli, Palestinian, and international water experts and policymakers.

In the fourth section of this chapter, water scarcity is discussed in order to reclaim it as a concept that is and should be grounded in daily, lived experience, rather than in technical rhetoric.

Water balance of Palestine

The aim of this section is to provide as accurate an image as possible of Palestine's water balance while taking into account various technical and political uncertainties. One conclusion emerges as especially significant from this discussion: whereas Israeli experts tend to estimate shared water resources at a lower level

---

[2] For an overview of the scholarship produced in the field of STS, see Jasanoff et al. (1994) and Jasanoff and Wynne (1998). For more on policy and political studies of science, see Ezrahi (1990) and Jasanoff (1991 and 1995).

than their Palestinian counterparts, they tend to estimate exclusively Palestinian water resources at a higher level than Palestinian experts.

The reason for this difference is predominantly cultural. On the one hand, Israeli water experts function within a generalized conception of "scarcity" inherited from the early years of the state: scarcity of water, scarcity of security, scarcity of land, and scarcity of population.[3] Since then scarcity, including water scarcity, became more than a technical portrayal of natural conditions; it evolved into both a dominant perception of Israeli experience and an active determinant of Israeli identity. Consequently, scarcity, no matter how imaginary, became a source of legitimacy for many Israeli policies, especially those related to the management of water resources.[4] We still witness the residual effects of that culture of scarcity: almost all Palestinian water resources shared with Israel are generally perceived as scarce resources that should be protected against new claims; conversely, exclusively Palestinian resources, such as the eastern groundwater basin or seasonal wadis, are perceived as abundant or as being at least potentially usable. A corollary perception to scarcity is the belief, especially prevalent during the 1950s and 1960s, that the "other," in this case the Arab states or Palestinians, live under conditions of abundance: an abundance of land, population, water, and security. Although one can hardly affirm any form of abundance when discussing contemporary Palestinian conditions, as we shall see here, the culture of scarcity still manifests itself, albeit in different forms.

On the other hand, Palestinian experts, especially before 1994, depended on Israeli sources for estimates of water balances. Their experience under military occupation shaped their suspicion that any and all Israeli estimates were politically motivated. Thus, Palestinian researchers tend to estimate the potential of shared water resources at higher levels than their Israeli counterparts and to estimate exclusively Palestinian water resources at lower levels.

These diverse perceptions of water estimates, induced by differences in the political culture of expertise, are expected to determine the distribution regime of shared water resources in the Final Status Talks of the peace process. The lower the estimates of shared water resources, the lower the Palestinian share of these

---

[3] The coemergence of scarcity and the state of Israel is discussed in detail in Alatout (1998b).

[4] On water scarcity as a source of legitimacy for Israeli water policies, see Alatout (1998b).

resources is expected to be; conversely, the lower the estimates of Palestinian resources, the more Palestinians will demand from these shared water resources.

With these general observations in mind and to provide a tenable picture of water balances in Palestine, this chapter will present Palestine's water resources in the form of ranges from least to highest estimates. However, in the next section, on the politics of numbers, two specific examples will be cited to demonstrate the ways in which the numbers used in technical debates about Palestine's water balance reflect their political contexts.

Rainfall in Palestine varies temporally and spatially. It can be as low as 150 mm/year in the eastern parts of Central and Southern Palestine and as high as 1 100 mm/year in the northern and mountainous parts of Central Palestine.[5] Overall average annual rainfall was recently estimated at 409 mm in Central Palestine and at 275 mm in the south. This translates into a volume of 2 349 Mm³, of which only about 100 Mm³ is in the southern parts of Palestine. Although most of this water is lost to evapotranspiration and surface runoff, an estimated 550 to 700 Mm³ percolates into groundwater aquifers.[6]

Palestine's water resources are, for the most part, shared with other states in the region. The next section is a description of the main water resources to which Palestine has a claim. Figure 1 shows that these resources are divided into two main categories: (1) surface water (including the Jordan River and seasonal wadis); and (2) groundwater aquifers (including the eastern, western, and northeastern basins).

---

[5] Anonymous source, Palestine Water Authority, interview, 20 July 1997. The place nomenclature used in this paper needs to be clarified before we go any further. Instead of the customary "West Bank" and "Gaza Strip," "Central Palestine" and "Southern Palestine" will be used, respectively. The reason for this choice is that the former terms are the discursive manifestation of the multiple occupations that Palestine has suffered. Although the decision to use these terms is mine, I am indebted to David B. Brooks for pointing me toward critically examining the naming of geographic spaces.

[6] Estimates are found to differ, depending on the research study. One estimate runs as high as 2 800 Mm³/year. The same source estimates water that percolates to underground aquifers at 625 Mm³/year. See United Nations (1991).

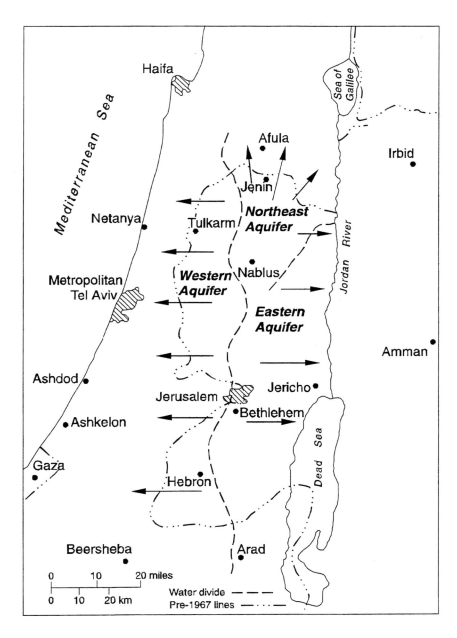

Figure 1. Palestine water resources.

Surface water

The Jordan River system

In addition to Palestine, there are four other riparians to the Jordan River: Israel, Jordan, Lebanon, and Syria. Although no formal water-sharing regime for the basin exists, Eric Johnston's Unified Plan functioned as the most authoritative sharing regime between 1956 and 1967. The Unified Plan was not ratified by the political institutions of the various countries involved, although it was endorsed by technical committees in the Arab states and Israel, and, on this basis, it functioned as a yardstick for the distribution of the Jordan River basin waters until the war of 1967. So long as Central Palestine was under Jordanian rule, the Unified Plan envisioned a West Ghor Canal that would divert water from the Jordan to irrigate farmlands along the western bank of the river, but experts disagree as to the quantity of water that was meant for the West Ghor Canal. Estimates range between zero, as asserted by some Israeli and Jordanian government officials, and 120 to 220 Mm^3/year (Table 1), as asserted by various Israeli and Palestinian water experts.[7] After the war of 1967, Palestinians were not allowed to use their share of Jordan River waters, despite the fact that Israel increased its use of the river basin, presumably for Central Palestine's population needs.

Seasonal wadis

Seasonal wadis are streams that suddenly come into existence as a result of winter rainstorms in which high precipitation accumulates in large volumes in a short time, 50 mm in 1 day or 70 mm in 2 consecutive days (Al-Khatib and Assaf 1992). This water resource is currently underused. It is estimated that these wadis can yield 66 to 99 Mm^3/year, 2 to 25 of which is received in Southern Palestine.[8] Whereas Palestinian water experts tend to think of this resource as less important

---

[7] The Jordanian rejection of the Palestinian claim was mentioned by an anonymous Jordanian water expert (interview, 3 September 1998). The same was asserted by the Israeli Water Commissioner, Meir Ben Meir (interview, 14 August 1997).

[8] Anonymous source, Palestine Water Authority, interview, 5 September 1998. For the lower estimates, see Al-Khatib and Assaf (1992).

than groundwater resources, Israeli experts tend to focus attention on the importance of using seasonal wadis.[9]

Groundwater aquifers

Groundwater aquifers constitute the main water resource of Palestine. They vary in depth from the Pleistocene gravel, tens of metres deep, to the Lower Cenomanian, hundreds of metres below the earth's surface (Issar 1990). Four basin aquifers are important, three of which are in Central Palestine (the western, northeastern, and eastern aquifers) and one of which is in Southern Palestine and is part

Table 1. Water resources by source and relative consumption (Mm^3).

Water resource to which Palestinians have a claim	Total resource	Used in Israel	Used in settlements	Used in Central Palestine	Used in Southern Palestine	Unused
Western aquifer [a]	310–362	313–333		10	21–27	
Eastern aquifer [b]	80–172		35–100	62–78		58
Northeastern aquifer [c]	131–145	101–115	5	20–25		
Gaza aquifer [d]	60–79	240–300				
Jordan and Yarmuk rivers [e]	1 060–1 287	560–650 (120–220 Palestine)				
Total	1 666–2 045	1 219–1 403	56	103–152 (includes springs)	73	

Source: Compiled by author.
Note: Excluding the eastern aquifer (all within Central Palestine), total water to which Palestine has a claim is 1 586–1 873 Mm^3/year, of which about 125 Mm^3/year is currently used by Palestine.
[a] Al-Khatib and Assaf (1992), Beschorner (1992), Assaf et al. (1993), Lowi (1993), and Sabbah and Isaac (1995).
[b] Al-Khatib and Assaf (1992), Isaac (1992), Assaf et al. (1993), Lowi (1993), and Assaf et al. (1994).
[c] Assaf et al. (1993) and ARIJ (1997).
[d] Beschorner (1992) and Isaac (1994).
[e] Beschorner (1992), Soffer (1992), Assaf et al. (1993), and Sabbah and Isaac (1995).

---

[9] Jacobo Sack, Senior Water Engineer of Mekoroth, stressed the importance of using these resources (interview, 21 July 1997).

of the Israeli coastal aquifer. Water from these aquifers is pumped through more than 370 wells in Central Palestine (close to 340 controlled by Palestinians and 36 controlled by Mekorot, the Israeli Water Company) and about 250 wells in Southern Palestine. Most of the Palestinian wells were drilled before the Israeli occupation started in 1967. Between 1967 and 1989, Palestinians were not granted permission to drill new wells or to maintain existing ones. In the mean time, Mekorot drilled the already mentioned 36 new wells to supply Israeli settlements in Central Palestine with water for irrigation and domestic use (Lowi 1993).

A combination of Israeli policies had three interconnected results: first, although Palestinian wells rarely reached a depth of more than 100 m, Mekorot's wells were normally sunk to a depth of from 200 to 750 m; second, whereas an Israeli well in Central Palestine yields, on average, 1 Mm^3/year, its Palestinian counterpart yields only 150 000 m^3/year; and third, the total yield of Palestinian wells was kept at its pre-1967 level of about 36 Mm^3/year.[10]

The western basin

The western basin is the largest in the formation. Although more than 80% of its recharge area lies in Central Palestine, 80% of the basin itself lies within the Green Line (Israeli boundaries before the 1967 war) and flows westward into Israeli territory. Estimates of the western basin's annual renewable yield vary between 310 and 362 Mm^3 (see Table 1). This variation in estimates is a result of the politics inherent to technical language: Al-Khatib and Assaf (1992) estimated the water potential at 335 Mm^3; Assaf et al. (1993), at 310 Mm^3; Shuval (1997[11]), at 310 Mm^3; Lowi (1993), at 360 Mm^3; ARIJ (1997), at 350 Mm^3; and a source from the Palestinian Water Authority recently estimated 362 Mm^3.[12] It is important to notice that although Al-Khatib and Assaf (1992) gave the middle estimate of 335, the same researchers, when collaborating with Israeli experts in Assaf et al. (1993), would later give a lower estimate of 310. This is a specific

---

[10] This figure only includes water pumped from wells. It does not include water from springs or water bought from Mekorot.

[11] Interview, 28 July 1997.

[12] Anonymous source, Palestine Water Authority, interview, 5 September 1998.

example of the assertion, made earlier, that estimates given by Israeli researchers are generally lower than those given by Palestinian experts for shared water sources, and this can be explained by the cultural tendencies mentioned above. To reiterate, whereas Israeli researchers tend to estimate shared groundwater potential at lower levels than their Palestinian counterparts, Palestinian experts tend to do the opposite. Although most of the water from this basin, 313 to 333 Mm^3, is used by Israel within the Green Line, 10 Mm^3 is used by Israeli settlements in Central Palestine. In other words, more than 90% of the western aquifer is used by the Israeli population. Only between 21 and 27 Mm^3 is used by Palestinians living in Central Palestine. The issue of sharing the western basin is expected to be one of the most salient in the Final Status Talks on water.

The northeastern basin

The northeastern basin contains the second largest aquifer in Central Palestine; it yields between 131 and 145 Mm^3/year (see Table 1), depending on the researcher. Most of this basin's water originates from rainfall in Central Palestine and flows toward the north and northeast into Israeli territory within the Green Line. About 75% of it is used by Israelis within the Green Line (101 to 115 Mm^3). Palestinians in Central Palestine use 20 to 25 Mm^3/year, and settlers in the same region use about 5 Mm^3/year from this basin. Again, as in the case of the western basin, the issues of whether or not to share this resource and how much to share are expected to become important in the Final Status Talks.

The eastern basin

The eastern basin is not an international body of water; the whole formation lies within the boundaries of Central Palestine. Water in this basin flows eastward and discharges into the Jordan River. Estimates of its potential vary considerably, between 80 and 172 Mm^3/year (see Table 1). Here we see the converse of the phenomenon described above: whereas Israeli experts tend to prefer higher estimates, their Palestinian counterparts prefer lower ones, as an exclusively Palestinian resource is at issue. However, many argue that the reason for the large variation also lies in the fact that much of this basin's water is saline and thus excluded from some calculations. Most of the water from the eastern basin is used by Palestinians in Central Palestine, 62–78 Mm^3/year, and a substantial portion is used in Israeli settlements, 35–50 Mm^3/year.

Gaza Basin

The Gaza Basin in Southern Palestine yields between 60 and 79 Mm^3/year (see Table 1). Of the 11 Mm^3/year by Israelis, 5 Mm^3/year is used within the Green Line, and 6 Mm^3/year is used in settlements throughout the south. The water in this basin is the most polluted in the area. Some experts estimate the salinity of this water to be higher than 2 000 mg/L in some areas, and it is increasing by 15 to 20 mg/L per year.

Summary of water balance

As shown in Table 1, of a total of 1 586 to 1 873 Mm^3/year of shared water (which excludes the eastern aquifer), Palestine gets a mere 125 Mm^3/year. This is equivalent to 6–8%. Israel (including the settlements) uses 70–77% of the same resource. Table 2 shows estimates of Palestinian water consumption from all resources, including the eastern aquifer, to be 176 to 225 Mm^3/year, which amounts to 88 to 112.5 m^3/person per year for all purposes.[13] Table 3 shows the detailed consumption of water divided into sectors and projected to 2005. The assumption is that the population in 1990 was 2 million people and that by 2005, it will be 3 million. As can be seen in Table 3, Awartani (1992) projected a substantial increase in water demand for household and agricultural uses. Of a projected water demand of 178 m^3/person per year, Awartani saw a need for 75 m^3/year of household water, more than doubling the 1990 pattern of per capita consumption, estimated at 31 m^3/year. Some 80 m^3/year was Awartani's projected per capita water demand for agriculture, a slight increase from the 1990 consumption, which was estimated at 77 m^3/year.

To show the significance of these numbers, Table 4 provides a comparison of Palestinian consumption with that of the Israeli and settler populations. Most experts believe that a settler uses as much as five times what a Palestinian uses.[14] But Israeli per capita consumption of water for household use, including that of

---

[13] These estimates assume a population of 2 million.

[14] Beschorner (1992) mentioned some estimates that settlers' per capita consumption of water is as high as 17 times that of Palestinians. However, along with Beschorner and an anonymous source in the Palestine Water Authority, this researcher prefers the more conservative figure. The exaggerated figure is mentioned here to demonstrate the large differences in estimates when it comes to water consumption patterns.

Table 2. Water consumption by region in Palestine (Mm^3/year).

Central Palestine total consumption	Southern Palestine total consumption	Palestinian total consumption
100–118	76–110	176–225

Source: Kuttab and Isaac (1993), Lowi (1993), and Planet (1998).
Note: This translates into an individual consumption of 88–112.5 m^3/person per year.

Table 3. Water consumption by sector in Palestine (Mm^3/year).

Year	Domestic	Industrial	Agricultural	Loss	Total
1990 [a]	64.4	7	154	45.3	271
2005 [a]	223.4	29.3	244	39.7	536.4
1990–92 [b]	31–53		80–162		111–215

Note: This amounts to the following detailed breakdown in Mm^3/person per year:

	1990–92	2005
Domestic	15–32.2	75
Industrial	3.5	10
Agriculture	40–81	80

[a] Awartani (1992).
[b] Al-Khatib and Assaf (1992) and Beschorner (1992).

Israeli citizens within the Green Line, is estimated to be at least three and maybe as many as five times that of Palestinians. If an equal per capita consumption of water were assumed, the Palestinian population would need 752 Mm^3/year, rather than the 225 Mm^3/year made available at present.

Table 5 shows the projected Palestinian population and water demand for various sectors. When assuming a doubling in per capita consumption, from about 110 m^3/year to 200 m^3/year, Palestinians will need 1 002 Mm^3/year by 2020, at a population level of 5 million. The fact that only 225 Mm^3/year is available at present calls for immediate action. Estimates can run even higher, and as shown in Table 6, demand might reach 1 263 Mm^3/year.

Table 4. Per capita use of water in Palestine (m³/person per year).

Israel (domestic)	Settlers in Central Palestine	Settlers in Southern Palestine	Palestinians (domestic)	United States
350–376 (103) [a,b]	354–591 [a,c]	2 326 [a,d]	119–133 (31–37) [a,b]	2 500 [e]

Note: It is generally agreed that per capita consumption by Palestinians is about one-fifth to one-third that of settlers.
[a] Beschorner (1992).
[b] Awartani (1992).
[c] Planet (1998).
[d] Isaac (1994). The distribution was as follows: Israelis, 375 m³/year; and Palestinians, 125 m³/year. A Gaza settler uses 2 000 m³/year.
[e] Naff (1992).

Table 5. Projected growth of Palestinian water demand, 1990–2020 (Mm³/year).

	West Bank	Gaza Strip	Total
Household			
1990	53	29	82
2000	115	62	177
2010	209	113	322
2020	340	186	526
Agriculture			
1990	70	70	140
2000	146	70	216
2010	234	70	304
2020	345	70	415
Industrial			
1990	5	2	7
2000	13	5	18
2010	27	11	28
2010	44	17	61

Source: Reproduced from Sabbah and Isaac (1995).

Table 6. Palestinian population projections and water consumption patterns by sector (Mm^3/year).

Year (population in millions)	Domestic	Agricultural	Industrial	Total
1990 (2)	78	140	7	225
2000 (3.45)	263	217	18	495
2010 (4.78)	484	305	37	826
2020 (6.22)	787	415	61	1 263

Source: Reproduced from ARIJ (1997).

Technical expertise and the politics of numbers

In the last section, it was demonstrated that estimates of water potential in Palestine vary greatly between Israeli and Palestinian water experts. (Differences also occur among Palestinian experts.) Explanations for the discrepancies are interesting in their own right. Israeli experts often assume that they follow rigorous scientific methodologies in investigating water potentials and that Palestinian experts are captured by political interests or, worse yet, scientifically unqualified to undertake such research.[15] The same is asserted by Palestinian water experts, only in reverse: Israeli experts are perceived as being captured by the political interests of occupation, whereas Palestinian researchers are perceived as the defenders of scientific integrity untainted by political maneuvering. This boundary work of identifying scientifically valid, as opposed to politically motivated, knowledge is part of the struggle to establish scientific legitimacy that, in turn, translates into a political resource.[16]

Research demonstrates, however, that conspiratorial perceptions of scientific expertise rarely hold true, except in extreme and limited cases where, for example, the policy-making apparatus is dealing with highly sensitive information.

---

[15] These sentiments were expressed to this researcher by many Israeli experts during interviews in 1996–97.

[16] Some early work in STS demonstrates the ways in which experts engage in boundary-defining practices aimed at shaping an expert community, its membership, and its legitimate ways of knowing. See Gieryn (1983) and Jasanoff (1991).

In general, research demonstrates that politics shapes the content of knowledge about water only as part of the cultural context in which that knowledge is produced. This is not a specific character of knowledge about water, nor is it specific to the Middle East; it has relevance to a broader discussion in the field of STS about the relationship between scientific practice and political context.

The academic field of STS emerged in the late 1980s from an earlier genre of scholarship on the politics, policy, history, sociology, and philosophy of science and technology.[17] Two empirical insights from STS are especially relevant to our purposes here. First, scientific knowledge is socially constructed. In this view, it is representative of more than just the natural order; it is also representative of the cultural, institutional, and historical conditions of its production. What we take as scientific fact is partly shaped by the sociopolitical context in which it is generated. The second relevant insight of STS is that scientific knowledge is not simply a representation of the natural and the sociopolitical; it is also a prescription, an attempt to define and shape both the natural and the sociopolitical orders.

Accepting the fact that technical knowledge about water is politically embedded, the question is whether we can identify and unpack the political content of that knowledge. This exercise is potentially empowering because it allows for the conscious redefinition of that unconscious political content, a chance to limit its repressive possibilities while strengthening its emancipatory potential. This will be illustrated by two examples from debates over Palestine's water balance.

Although a politics of rights is embedded in the technical language of both Israelis and Palestinians, Palestinian experts are generally more forthcoming in recognizing this. Israeli experts, for the most part, reject the focus on the political language of rights as a waste of time. They prefer a technical language that sidesteps politics, assuming that the technical language of numbers speaks on its own in clear, convincing, and indisputable ways.

Focusing on a language of rights, Palestinians insist on addressing the issue of the equitable redistribution of shared water resources, one that grants to Palestinians riparian rights over known resources (Awartani 1992; Isaac 1992, 1994; Assaf 1994). Palestinians point mainly to the unfair distribution of shared water

---

[17] To trace the evolution of STS, compare these texts from the 1980s and 1990s: Barnes and Edge (1982), Collins and Pinch (1982), Jasanoff (1991), Knorr-Cetina (1982), Knorr-Cetina and Mulkay (1983), Latour (1993), Latour and Wolgar (1986), and Pinch (1986).

resources: since 1967 Palestinians did not receive their share of the Jordan River resource; more than 85% of the western aquifer is used by Israelis, including the settler population; of the northeastern aquifer, more than 75% is used by the Israelis. For the most part, the political meaning of these numbers is explicitly addressed by Palestinian researchers.

Israeli experts differ, depending on their relation to the water-policy apparatus. Government officials reject out of hand the question of these rights. For example, to explain the differences in per capita water consumption, the Israeli Water Commissioner, Meir Ben Meir, pointed to "cultural differences" and differences in "standards of living."[18] The Minister of National Infrastructure, Ariel Sharon, dismissed any Palestinian focus on water shortages as political, as "another excuse to suspend the [diplomatic] process and the talks, through exploiting a situation of distress that stems both from their own deficient methods and from the recent severe heat wave" (Hass 1998c). The Minister of Agriculture, Rafael Eitan, sees all this Palestinian focus on water shortage as mere lies. When asked about the Palestinian water-shortage stories, he made the following statement to a National Public Radio reporter: "You are perhaps a naive person who has been sucked into a culture that is maybe based on lies and deception" (NPR 1997). Israeli experts with official affiliation in the government argue, as did Meir Ben Meir at one point, that there are no Palestinian rights to water over and beyond what Israel provides, that is, 225 Mm^3/year, especially for irrigation. Indeed, if there is a need for more water, the argument goes, then Palestinians have to search for other resources somewhere else.[19]

Despite the official Israeli line, and somewhat surprisingly, some Israeli academics and peace activists do not seem to appreciate the positions taken by Palestinian water experts either. They sidestep the politics of rights by both denouncing it as divisive and stressing the importance of objective, nonpolitical, technical language. Gershon Baskin asserted that the Palestinian focus on the language of rights is "stupid" (Horan 1998). He went on to say that "the Palestinians are shooting themselves in the foot by concentrating only on water rights." The

---

[18] Interview, 14 August 1997. Also see Hass (1998c).

[19] Meir Ben Meir, interview, 14 August 1997.

same attitude was expressed by Hillel Shuval, although in a softer tone.[20] What Israeli experts fail to note is the fact that the technical language they deploy camouflages a political language of rights. The role of experts interested in working out permanent and peaceful solutions to water conflicts is to make explicit such politics.

A careful investigation of the numbers in Table 7, reproduced from a work by Hillel Shuval (1993), demonstrates this point clearly.[21] Shuval, in projecting water needs to 2023, estimated Palestinian water resources at 200 Mm^3/year and Israeli water resources at 1 500 Mm^3/year and concluded that by this year Israel would have an excess of 250 Mm^3/year and Palestinians would have a deficit of 425 Mm^3/year. What needs to be clarified is the fact that these calculations are true only if it is assumed that the present distribution of shared water resources is also the rightful one. Thus, Shuval's work rests on an unconscious concealment of the politics of rights, inadvertently hiding them behind the technical language of numbers. In the process, present distribution of shared water resources between Israel and Palestine is legitimized, even though the current situation was reached through the deployment of a variety of repressive policies during the Israeli occupation from 1967 to 1994, or even before that, in the denial of all Palestinian national rights since 1948.

Of course, the same technical language and its politics find support in the international legal principle that historic uses of a shared water resource should be respected. However, it assumes that this is the only relevant principle operative in the Palestinian–Israeli water conflict. The equally important international legal principle of the equitable use of a shared water resource, which was meant to be a check on the concept of historic right, is ignored.[22]

Another example of the politics of numbers is even more interesting because it comes from a Palestinian–Israeli collaborative project. Assaf et al.

---

[20] Interview, 30 July 1997.

[21] See Shuval (1993). In interviews with Shuval, he seemed much more aware of Palestinian rights than some other Israeli experts, on the left or the right of the political spectrum. Using an example from the work of this fair and open-minded water expert is done on purpose to show that no matter how inadvertently, a politics of rights infuses itself, necessarily and unconsciously, into technical language.

[22] See Ohlsson (1995). More on historic and equity rights will follow in the next section.

(1993) called for assigning a specific quantity of water as the minimum water requirement (MWR) that every person in Israel and Palestine would be entitled to. Per capita MWR was calculated at 125 m^3/year by the collaborative project. Others differed in their estimates. For example, Peter Gleick provided an influential estimate of between 75 and 150 m^3/person per year.[23] MWR is presumably a humane way to look at the issue, because it depoliticizes and humanizes water allocation by advocating the equal distribution of resources on the individual level. However, the application of MWR in this case is problematic: it envisions the allocation of all available water resources exclusively for household and industrial uses, and only until the year 2023, thus eliminating any meaningful role for agriculture in the process. The goal, arguably, will be achieved through water pricing and the elimination of water subsidies to agriculture.

Table 7. Population growth, water resources, and water balances in Israel, Jordan, and Palestine.

	Israel	Jordan	Palestine
Population (millions)			
1993	5	3	2
2023	10	10	5
Water-resource potential (Mm^3/year)	1 500	880	200
Cubic metres per person per year			
1993	300	250	100
2023	150	90	40
MWR in (Mm^3/year)			
2023	1 250	1 250	625
Net (Mm^3/year)	+250	−370	−425

Source: Reproduced from Shuval (1993).
Note: MWR, minimum water requirement.

---

[23] Shuval (1993) sets MWR at 125 cubic metres per person-year. The same is also assumed by Assaf et al. (1993).

A careful look at the statistics will demonstrate the hidden politics behind such a scheme. Whereas agriculture does not constitute an important activity in Israel, either in terms of gross national product (GNP) (2–3%) or in terms of labour (1.5–3%), it does contribute greatly to the Palestinian economy (25% of GNP and 30% of the work force). Seen from this perspective, this scheme is political and prescriptive through and through, one that envisions a purposeful alteration of a people's way of life, that of the Palestinians, without paying attention to local and cultural specificities. The collaborators on this project do not seek legitimacy for their vision through democratic processes, nor do they envision it as a political project at all. On the contrary, they perceive their numbers as technically legitimate, as a mere representation of water conditions, and they perceive themselves as the objective mouthpieces for nature. Eliminating agriculture is perceived as the consequence of such numbers, not as the keen desire of political agents, the experts themselves. Theirs is a naive and simplistic notion of scientific expertise. Scientific knowledge is at least as prescriptive as it is representative of nature, and in that sense it is strongly political.

There are two important reasons for unpacking the political meaning of technical language, especially in terms of water rights. First, there are financial reasons for assigning rights to water resources: the more of an entitlement a party has to existing water resources, the less that party has to spend on developing new ones. Second, there is the national–symbolic dimension: the national rights of Israelis and Palestinians were often negotiated through water politics. The Israeli state, in the 1950s and 1960s, used water not only in the battle to settle new immigrants in the Negev and elsewhere but also in the battle to achieve political legitimacy. Water was an issue in which Israeli identity took on a political and a geographic form (Alatout 1998a). Now, water is playing the same role again: it is an issue in which the legitimacy of contested national sovereignties is negotiated. The technical language of water potential cannot but be embedded in political meaning.

International law and water in the Middle East

Presumably, Israeli rights to water from the northeastern and western aquifers are based on the fact that these water sources were used during the 1930s through the 1950s by Zionist and then Israeli farmers. This establishes Israel's use of these aquifers as historic. The significance of this argument lies in the fact that the International Law Commission uses historic use as the basis for the assignment of rights among partners to a nonnavigational watercourse. To protect downstream countries — such as Israel, in the case of Central Palestine's aquifers — the International Law Commission calls for the protection of existing uses from "appreciable harm" that could result from new uses in upstream communities. However, Ohlsson (1995) suggested that the indiscriminate application of this principle results in many inequitable situations in which an upstream party is denied access to its water resources and thus to development. Article IV of the Helsinki Rules of 1966 addresses this problem specifically. It says that "each basin state is entitled, within its territory, to a reasonable and equitable share in the beneficial uses of the water on an international drainage basin."[24] This right to equitable sharing is usually argued in support of the Palestinian case for using the western and northeastern aquifers.

Two important comments need to be made on this. First, the legitimacy of Israel's historic right to the aquifers would be severely diminished by consideration that the Palestinian Declaration of Independence, issued in Algiers in 1988, was based on Security Council Decision 181 of 1947. The fact that the Palestinian state was made impossible all these years by the use of military force delegitimizes any arrangements manufactured after 1947. The only way to grant legitimacy, finally, to any post-1947 uses of natural resources would be through a treaty, signed and agreed on by the two parties. Second, even if one takes Israeli historic right to be legitimate, one cannot avoid the question of equity in the use of shared aquifers, a concept that is gaining momentum in international law.

Benvenisti (1986) rightly commented that the whole problem lies in balancing the rights stemming from historic use against those stemming from considerations of equitable sharing. Even given the fact that equity rights are vague and open to diverse interpretations, they raise the following questions: How can we simultaneously operationalize the principles of historic and equity rights? What does their application mean in terms of specific allocations of specific watercourses? The articles of the International Law Commission (ILC n.d.) provide some direction. Article 5 of the draft resolution, adopted by the United Nations

---

[24] Quoted in Ohlsson (1995).

General Assembly in 1997, stresses the importance of using an international watercourse "in an equitable and reasonable manner." Article 6 elaborates on the definition of "equitable and reasonable manner" as the manner in which the following factors are considered:

- Geographic, hydrographic, climatic, ecological, and other factors of a "natural" character;

- The social and economic needs of the states concerned;

- The effect of the use or uses of the watercourse by a state on other watercourse states;

- Existing and potential uses of the watercourse;

- Conservation, protection, development, and economy of use of the water resources of the watercourse and the costs of measures taken to that effect; and

- The availability of alternatives of corresponding value to a particular planned or existing use.

Even though experts will undoubtedly differ on the relative importance of each of these factors, the articles of the International Law Commission provide an acceptable basis for negotiation. They constitute an important invitation to policymakers and academics alike to think about water-sharing from different riparian positions.[25]

In conclusion, the relative weight of these factors cannot and need not be determined a priori and at a distance. Even the relevance of the various factors is,

---

[25] Operationalizing these factors was attempted by James Moore (1994), who assigned a different weight to each of the factors and then calculated the Israeli, Jordanian, Lebanese, and Palestinian shares of water resources. The main problem with Moore's resolution is that it deploys a seemingly neutral language of numbers that obscures the need for the political process of negotiation to determine the weights assigned.

itself, open to political contestation and should be determined through negotiations. What is important, however, is to simultaneously ground any possible resolution to the problem on the international legal principles of equity and historic right.

Scarcity, numbers, and experience

Article 40 of the interim agreement between the Israelis and Palestinians asserts that "Israel recognizes the Palestinian water rights in the West Bank. These will be negotiated in the permanent status negotiations and settled in the Permanent Status Agreement relating to various water resources."[26] Recognizing the impossibility of keeping Palestinian water allocation at the same level of 225 Mm^3/year, which had gone unchanged since 1967, the agreement stipulated that Israel would provide an additional 70–80 Mm^3/year in order to satisfy "future Palestinian needs." This was meant to take care of the urgent and immediate needs of Palestinians within the 5-year interim period. However, of these 70–80 Mm^3/year that Israel was supposed to provide, only 28.6 Mm^3/year has been received by Palestinians, and Palestine, moreover, was allowed to extract this quantity of water from the eastern aquifer, over which Israel has no claim. In other words, Israel has yet to fulfill its immediate commitment. This act of omission is not at all confidence inspiring.

During the summer of 1998, drought conditions in Palestine were severe. Hebron and Bethlehem were hit the hardest: some areas did not have running water for 2 months at a time. From 15 August, the Israeli media started paying attention to the lack of water for Palestinians. One Israeli reporter (Jehl 1998) expressed astonishment that although Hebron, a city of 200 000, had been denied water for a month, "there is no sign of a water shortage in the Jewish settlements just outside of Hebron. There and in Israel as a whole, residents still water lawns and wash their cars." Another Israeli journalist (Hass 1998a), talked about how the expected per capita supply of 30 m^3 fell to 6.7 m^3 for individuals making up a population of half a million in the West Bank (Central Palestine) during those 3 summer months. Isa Atallah, head of the Palestine Water Authority in Hebron, described the experience from the Palestinian point of view: "It is really frustrating

---

[26] The Israel–Palestinian Interim Agreement on the West Bank and Gaza Strip was signed in Washington, DC, on 28 September 1994.

when your children are going thirsty and you see the settlers next door watering their gardens and swimming in their pools" (Hass 1998b).

Water is scarce in the region — that is what many experts assert. They believe it, and they try to convince everyone of its truth. They spend much of their resources trying to establish water scarcity as an indisputable scientific matter of fact. Experts argue that water scarcity touches everybody in the region and that it is a shared environment; they hope that its harshness might induce the two nations to cooperate. All this could be appreciated if not for the ungrounded technicalization of the discourse of scarcity: the most circulated definition of water scarcity is a condition in which water supply is less than 500 m^3/person per year. This technical assessment of scarcity, however, is inadequate at best because it ignores the fact that within that technical limit, the lived experience of people might and does in fact differ along national–political lines (Israelis have access to substantially more water than Palestinians) and along social-class lines as well (Palestinian rural areas are hit hardest).

Scarcity cannot be solely about numbers, no matter how compelling, surprising, or shocking they may be. In both Israel and the settlements, water is flowing without limits. The Israeli experience of scarcity, grounded in numbers, is radically different from that of Palestinians, grounded in daily shortages and interruptions. In fact, to the Palestinian neighbour, the settlement looks very much like the land of abundance that lived for so long in the Zionist imagination.

Conclusion

Water scarcity is real in Palestine. It is not merely the conclusion of a theoretic calculation that constructs a scarcity threshold, usually set at 500 m^3/person per year, and then says that that threshold is not met. Rather, it is known through real experience and in practice. It has a human face that is demonstratively suffering and demonstratively Palestinian.

Although water scarcity is real in Palestine, it has not been induced by natural conditions alone. The numbers in circulation on water balances in Palestine reflect and, however inadvertently, seek to legitimize and perpetuate a given political and cultural order. This order functions through an imposed administrative arrangement in which some — Israeli settlers in Palestine and Israeli citizens within the Green Line — can and do use groundwater resources to satisfy their needs while others — the Palestinian population — have only limited and very unsatisfactory access to those resources.

The fact that politics is one of the main determinants of Palestine's water balance should not cause us to despair, as there is always an opportunity for political maneuvering. All that is needed, however difficult it may be to achieve, is to consciously redefine the political meaning of technical language and construct a technical discourse that somehow merges Israeli and Palestinian water interests.[27] A workable resolution to water conflicts between Israel and Palestine can only be achieved through negotiations that continuously work out emerging problems and critiques.

References

Alatout, S. 1998a. From water abundance to water scarcity: the emergence of an Israeli identity. Paper presented at the Annual Meeting of the Society of Social Studies of Science, 29 Oct 1998, Halifax, NS, Canada.

———— 1998b. "States" of scarcity: water, knowledge, and governing in Israel. Paper presented at the Annual Meeting of the Middle East Studies Association, 6 Dec 1998, Chicago, IL, USA.

Al-Khatib, N.; Assaf, K. 1992. Palestinian water supplies and demands. Paper presented at Water and Peace in the Middle East: the First Israeli–Palestinian International Academic Conference on Water, 1992, Zurich, Switzerland.

ARIJ (Applied Research Institute-Jerusalem). 1997. The Water Conflicts in the Middle East from a Palestinian Perspective. ARIJ, Jerusalem, Israel.

Assaf, K.; Al-Khatib, N.; Kally, E.; Shuval, H. 1993. A proposal for the development of a Regional Water Master Plan. Israel/Palestine Center for Research and Information, Jerusalem, Israel.

———— 1994. Water — a national resource for Palestinians — and the significance of its quality. Paper presented at Our Shared Environment: the First Israeli, Palestinian, and International Conference on the Environmental Challenges Facing Israel, the West Bank and Gaza, 1994, Tantur Ecumenical Institute, Jerusalem, Israel.

---

[27] Despite the critical comments in this paper on the substance of Assaf et al. (1993), for ignoring the importance of agriculture in Palestinian life, the work cited remains one of the rare attempts at redefining the terrain of water politics along the lines envisioned in this chapter.

Awartani, H. 1992. A projection of the demand for water in the West Bank and Gaza Strip, 1992–2005. Paper presented at Water and Peace in the Middle East: the First Israeli–Palestinian International Academic Conference on Water, 1992, Zurich, Switzerland.

Barnes, B.; Edge, D., ed. 1982. Science in context: reading in the sociology of science. MIT Press, Cambridge, MA, USA.

Benvenisti, M. 1986. Demographic, economic, legal, social and political development in the West Bank. American Enterprise Institute, the West Bank Data Base Project, Jerusalem, Israel.

Beschorner, N. 1992. Water instability in the Middle East. London, UK. Adelphi Report 273.

Collins, H.; Pinch, T. 1982. Frames of meaning: the social construction of extraordinary science. Routledge and Kegan Paul, London, UK.

Ezrahi, Y. 1990. The descent of Icarus: science and the transformation of contemporary democracy. Harvard University Press, Cambridge, MA, USA.

Gieryn, T. 1983. Boundary-work and the demarcation of science from non-science: strains and interests in professional ideologies of scientists. American Sociological Review, 48, 781–795.

Hass, A. 1998a. Dire water shortage in the West Bank. Haaretz, Jerusalem, 27 Jul 1998.

———— 1998b. Sharon says PA excuses are all wet. Haaretz (English edition), Jerusalem, 19 Aug 1998.

———— 1998c. Israel to ease West Bank water shortage. Haaretz (English edition), Jerusalem, 24 Aug 1998.

Horan, D. 1998. Palestinians, Israelis lock horns over water. *In* Website Dawn International: the Internet edition. wysiwyg://55/http://dawn.com/daily/19989323/int8.htm; cited 23 Mar 1998.

ILC (International Law Commission). n.d. Draft articles on the Law of the Non-navigational Uses of International Watercourses. http://allserv.rug.ac.be/~sdconinc/waternet/draftILC.htm

Isaac, J. 1992. Opening remarks. Paper presented at Water and Peace in the Middle East: the First Israeli–Palestinian International Academic Conference on Water, 1992, Zurich, Switzerland.

———— 1994. Sustainable development and the Palestinians. Paper presented at Our Shared Environment: the Proceedings of the First Israeli, Palestinian, and International

Conference on the Environmental Challenges Facing Israel, the West Bank and Gaza, Tantur Ecumenical Institute, 1994, Jerusalem, Israel.

Issar, A.S. 1990. Water shall flow from the rock: hydrology and climate in the lands of the Bible. Springer-Verlag, Heidelberg, Germany.

Jasanoff, S. 1991. The fifth branch: science advisers and policymakers. Harvard University Press, Cambridge, MA, USA.

———— 1992. Acceptable evidence in a pluralistic society. *In* Mayo, D.; Hollander, R., ed., Acceptable evidence. Oxford University Press, New York, NY, USA.

———— 1995. Science at the bar: law science and technology in America. Harvard University Press, Cambridge, MA, USA.

Jasanoff, S.; et al. Handbook of science and technology studies. Sage Publications, London, UK.

Jasanoff, S.; Wynne, B. 1998. Science and Decisionmaking. *In* Ryner, S.; Malone, E., ed., Human choice and climate change. Battelle Press, Columbus, OH, USA.

Jehl, D. 1998. Water divides haves from have-nots in the West Bank. The New York Times, 15 Aug 1998, p. A3.

Knorr-Cetina, K.D. 1982. The manufacture of knowledge: an essay on the constructivist and contextual nature of science. Pergamon Press, Oxford, UK.

Knorr-Cetina, K.D.; Mulkay, M., ed., 1983. Science observed: perspectives on the social study of science. Sage Publications, London, UK.

Kuttab, J.; Isaac, J. 1993. Approaches to the legal aspects of the conflict on water rights in Palestine/Israel. Applied Research Institute-Jerusalem, Jerusalem, Israel.

Latour, B. 1993. We have never been modern. Harvard University Press, Cambridge, MA, USA.

Latour, B.; Woolgar, S. 1986. Laboratory life: the construction of scientific facts (2nd ed.). Princeton University Press, Princeton, NJ, USA.

Lowi, M. 1993. Water and power: the politics of a scarce resource in the Jordan River basin. Cambridge University Press, Cambridge, UK.

Moore, J. 1994. Parting the waters: calculating Israeli and Palestinian entitlements to West Bank aquifers and the Jordan River basin. Middle East Policy 3(2), 91–108.

Naff, T. 1992. A case for demand-side water management. Paper presented at Water and Peace in the Middle East: the First Israeli–Palestinian International Academic Conference on Water, 1992, Zurich, Switzerland.

NPR (National Public Radio). Troubled waters. NPR. Interview, Nov 1997.

Ohlsson, L., ed. 1995. Hydropolitics: conflicts over water as a development constraint. Zed Books, London, UK.

Pinch, T. 1986. Confronting nature: the sociology of solar neutrino detection. Reidel, Dordrecht, Netherlands.

Planet. 1998. Palestine: the promising land. http://www.planet.com/inv/waterr.htm; cited 2 Oct 1998.

Sabbah, W.; Isaac, J. 1995. Towards a Palestinian water policy. Applied Research Institute-Jerusalem, Jerusalem, Israel.

Shuval, H. 1993. Approaches to finding an equitable solution to the water resources problems shared by Israelis and the Palestinians in the use of the mountain aquifer. *In* Baskin, G., ed., Water: conflict or cooperation, Israel/Palestine issues in conflict issues in cooperation. Israel/Palestine Center for Research and Information, Jerusalem, Israel. pp. 37–84.

Soffer, A. 1992. The relevance of the Johnston Plan to the reality of 1993 and beyond. Paper presented at Water and Peace in the Middle East: the First Israeli–Palestinian International Academic Conference on Water, 1992, Zurich, Switzerland.

United Nations. 1991. Report Prepared by the Economic and Social Commission for Western Asia on Israeli Land and Water Policies and Practices in the Occupied Palestinian and Other Arab Territories. United Nations, New York, NY, USA. A/46/263.

Chapter 5

EVALUATING WATER BALANCES IN JORDAN

Esam Shannag and Yasser Al-Adwan

Introduction

This chapter examines water balances in Jordan, with the view that water is a precious commodity on which human life depends and a limited resource of strategic importance for the coming years. It relies on the findings and conclusions of studies examined in a general review of literature on the topic.

Jordan's population currently exceeds 4.6 million (PRB 1998), and its annual rate of increase is 2.5%. It will take 28 years for Jordan to double its population, and by 2025, it is projected that its population will reach 10 million, barring any unforeseen circumstances, such as massive refugee in-migration as a result of political instability or armed conflict in neighbouring countries.

Jordan has endured deficits in water resources since the early 1960s. The country is classified as water scarce (Table 1), compared with countries in the region categorized as water stressed (for example, Cyprus and Egypt) or water abundant (for example, Lebanon and Syria).

Israel, Jordan, Lebanon, Palestine, and Syria are all considered riparian to the Jordan River system. In the eyes of many regional water experts, this makes the Jordan River basin the most likely flash point for conflict in the Middle East. It also creates a common area for cooperation between these same countries to resolve their water disputes or conflicts, bilaterally and multilaterally, as the Jordanian–Israeli Peace Treaty has recently illustrated.

This workshop, the interaction of its participants, and the publication of the papers presented here can make a contribution to regional cooperation by building

Table 1. Annual renewable water per person. [a]

	Renewable annual freshwater available per person (m^3/person per year)	Rank
Water-scarce countries (20 countries)		
Jordan	327	10
Israel	461	12
Water-stressed countries (8 countries)		
Egypt	1 123	23
Cyprus	1 282	25
Water-abundant countries		
Lebanon	1 818	31
Syria	2 087	36
Turkey	3 626	61
United States	9 913	91
Canada	108 900	141

[a] These are some of the 149 countries classified by water availability in 1990.

confidence among water researchers and policymakers in the region and by examining, analyzing, and recommending ways to reduce tensions in this region. We believe that the minimum water requirement (MWR) can be met for all involved and that this can be made the nucleus for peacemaking and future cooperation.

Supply of water in Jordan

Jordan's primary sources of water are aquifers and basins (Table 2) fed and recharged through annual rainfall. The Yarmouk Basin is the largest in the country. Water from ground, surface, and nontraditional sources (Table 3), exhibits short- and long-term variations, and this requires that decision-makers in charge of planning and development be informed and advised on the general and specific data. Jordan's water supply suffers because about 85% of the total amount of water is

Table 2. Aquifer and basin water status in Jordan (Mm³/year).

Basin	Used	Available
Yarmouk	59	40
Jordan River tributaries	6.3	15
Jordan River plains	21.7	21
Amman and Zarqa	153.8	87.5
Dead Sea	68.6	57
Diesa	56	100
North Wadi Araba	1.75	3.5
South Wadi Araba	4	5.5
Jaffar	23	27 (18)
Azraq	32	28
Sarhan	0.8	5
Hamad	1.8	8

Table 3. Water resources: ground, surface, and nontraditional (Mm³).

Renewable groundwater	280
Nonrenewable groundwater	118
Surface water	755
Treated wastewater	22

lost to evaporation annually, which leaves only a small amount of surface and groundwater to enter the water supply.

Many methods have been suggested to increase the water supply, including intensive capturing of rainwater through the use of micro- and macrodams, desalination of sea water, and importation of water from neighbouring countries, as well as other alternatives. However, all these are subject to cost–benefit analyses and geopolitical constraints.

Table 4. Water usage in Jordan, 1985–2005 (Mm^3).

Sector	1985	1989	1995	2005
Domestic	200	242	254	301
Agricultural	624	721	877	1 067

Table 5. Distribution of water use in Jordan (Mm^3).

Source	Total by source	Domestic and idustrial	Irrigation
Renewable groundwater	275	155	220
Nonrenewable groundwater	56	11	45
Surface water	530	30	500

Demand for water in Jordan

Water in Jordan is used primarily for agriculture (Table 4). Agriculture accounts for 77.5% of all water consumed, the rest being for domestic and industrial use. Annual growth in demand for water in Jordan is estimated at 25 Mm^3/year. This growth is related to urbanization and industrial expansion, as well as to increased domestic use, mainly as a result of population growth (Tables 5 and 6). End use by sector will be discussed in this chapter, but, at present, detailed data are unavailable.

The current situation of water supply and demand in Jordan raises serious concerns about the country's water balance, as well as about the qualitative deterioration of water. The picture is so gloomy that any water researcher would observe that it is all too easy for the country to "cross the red line" when faced with annual water deficits, overuse, resource depletion or contamination, and machine and human error — witness the case of water contamination in West Amman in 1998. Projections of water resources to 2025 (Table 7) demonstrate that there will be persistent shortage.

Table 6. Population versus per capita water availability.

Country	Total annual renewable fresh water available (Mm^3)	1955 population (millions)	1955 per capita water availability (m^3)	1990 population (millions)	1990 per capita water availability (m^3)	2025 population (millions)	2025 per capita water availability (m^3)
Canada	2 900 987	15.763	184 354	26.639	108 900	32.83	88 364
Cyprus	900	0.053	1 698	0.702	1 282	0.839	1 073
Egypt	58 874 000	22.99	2 561	52.426	1 123	86.483	681
Israel	2 418	1.748	1 229	4.66	461	7.318	294
Jordan	1 331	1.447	906	4.009	327	10.299	1 236
Lebanon	4 981	1.613	3 008	2.74	1 818	4.028	1 236
Syria	25 785	3.967	6 500	12.355	2 086	34.072	757
Turkey	203 023	23.859	8 509	55.991	3 626	84.537	2 401
United States	2 487 002	165.932	14 934	249.975	9 913	301.716	8 231

Table 7. Water resources in Jordan, 1985–2005 (Mm³).

	2005		1995		1989		1985	
	C	A	C	A	C	A	C	A
Renewable groundwater	390	280	359	280	375	280	313	280
Nonrenewable groundwater	118	118	118	118	56	118	25	118
Surface water	755	755	594	594	500	500	466	466
Treated wastewater	60	60	60	60	32	32	20	20

Note: A, available; C, consumed.

Several methods are in place to help alleviate the shortage, with reduced consumption at the top of the list. Appropriate pricing is a preferred alternative for achieving this goal. Money saved and funds generated may justify installing and using new technologies more efficient in terms of cost–benefit analyses.

The current system adopted by the Jordanian government is based on political considerations far removed from economic ones, and it is assumed that future governments in Jordan will not deviate from this policy. However, in our opinion, other alternatives for solving water-shortage problems now and in the future, based more on economic than political considerations, must be pursued and adopted.

The role of geopolitics

It is an established fact, recognized by the scientific community, that at any time consumption far exceeds production, a crisis is at hand; therefore, in Jordan, we can legitimately speak about a water crisis. This manifests itself at two different levels: the national level and the regional level, including Jordan and neighbouring countries, such as Egypt, Iraq, Israel, Lebanon, Palestine, Syria, and Turkey.

Although there has been some water cooperation, the other side of the coin is water conflict. This is reflected in bilateral competition for shared resources, and multilateral as well as regional tensions. This negative facet of the water issue has been ongoing and highly counterproductive for the past 50 years. The most realistic and hopeful option to be pursued is cooperation, and the safest way of cooperating is at the regional level, where bilateral agreements must be negotiated and reached and treaties must be signed. In this way, a solution can be found to the serious problems built up over the last few years around water shortages in relation to other geopolitical and multilateral issues, such as forced migrations and refugees. For Jordan in particular, the problem of water has been exacerbated by the influx and migration of displaced people coming into the country from neighbouring areas. Recently, there was a sudden 10% increase in Jordan's population, as a result of the Gulf War. Some 400 000 refugees fled from Iraq to Jordan and added their numbers to the large population of Egyptian workers already in Jordan for socioeconomic reasons.

MANAGING DEMAND — Because water supply has not kept pace with consumption, current water policy emphasizes the economic aspects of water demand. A key element in Jordan's current policy for managing the water problem is a move toward water-demand management. Analysis of economic sectors, the uses of water, and economic efficiency are only a few aspects of the economic management of water in the country. A pricing mechanism for household water consumption has recently been applied, whereby the price for every cubic metre over and above the first 20 m^3 in the water cycle has been increased. The objective of policymakers was the efficient use of water and the application of cost-effective options with the guarantee of a basic minimum per capita supply at an accessible price. Water from wells is rationed for agricultural consumption.

PUBLIC EDUCATION — Workshops and seminars have been held by various institutions, such as the Ministry of Water, the Jordan Valley Authority, and the universities, to discuss water problems, consumption behaviour, and water-saving mechanisms. The media (including the advertising media) has been employed to raise public awareness of the seriousness of the water shortage and related consumption problems. Furthermore, government agencies have also taken cultural values into account, holding public forums with a focus on the issue in relation to Islamic teachings.

SEARCHING FOR WATER — Several options can be pursued in the search for water to meet critical human needs. One is the Diesa Project. Diesa is a groundwater base shared by Jordan and Saudi Arabia. The cost of the project (pipelines, dam construction, and distribution) is estimated at $600 million United States dollars.

Another option is negotiating the future supply of water by pipeline from Iraq. The most dramatic option would be large-scale water imports from Turkey.

WATER SECURITY AND AVAILABILITY — Regional security and cooperation cannot be the concern only of one party or some parties in the region but must be the concern of all governments and people throughout the whole region, regardless of nationality. Therefore, advocating security as a factor for withholding water or for not cooperating with bordering nations may become the main reason for the defeat of the security argument itself. Everyone needs security, everyone draws a red line, and cooperation may be the only feasible solution to serve the interests of all people in the region, especially their security interests.

THE WATER SAFETY NET — Arguments on water rights fuel disagreements on transboundary water issues. Even though rights advocates may have some valid beliefs, in a turbulent region a safety net of water supply for all may be much more appropriate for enhancing cooperation and security. Therefore, just and fair allocation of water is a conceptual basis far more likely to work.

DEMAND MANAGEMENT IN A TRANSBOUNDARY CONTEXT — The primary regional objective of water demand management is not to deal with who rightfully owns the water but with how one can successfully manage transboundary water issues.

People are divided into three camps in the Middle East on the issue of how to deal with water policy:

- *The regional-cooperative-management camp* — The main focus of this camp is to look at demand requirements and transboundary water issues. It sees a political as well as a management resolution to the issues. A regional centre for conflict resolution would likely be of importance to working out, in part, resolutions between the conflicting views and focuses on water in the region.

- *The nationalist camp* — This camp postulates that controlling water heads, resource leads, and their flow and distribution is tied to sovereignty, and, as such, it is a national matter.

- *The economic-efficiency-allocation camp* — This camp advocates that an efficient pricing mechanism and an economic allocation model would work at the national but not at the transboundary level. However, water resource management at the regional level may also encompass efficient allocation of water resources.

Conclusion

If water is to be considered in terms of the criterion of economic allocation, a host of related economic issues must be dealt with, as integral parts of any economic model. Water consumption per person, labour in economic sectors, agriculture, per capita income, and economic disparities are issues to be considered and built into any economic model for efficient allocation, pricing, and consumption of water.

That being said, we believe that transboundary management of water is vital for cooperation and stability in the region (peaceful coexistence). Cooperation and stability must be the two pillars of enduring peace in the Middle East.

Reference

PRB (Population Reference Bureau). 1998. World population data sheet. Demographic data and estimates for the countries and regions of the world. PRB, Washington, DC, USA.

Chapter 6

TURKEY'S WATER POTENTIAL AND THE SOUTHEAST ANATOLIA PROJECT

Mehmet Tomanbay

Introduction

As one of the relatively water-rich countries in the Middle East, Turkey often finds itself in the midst of discussions at international meetings on water issues in the region. In these discussions, it is usually assumed that Turkey is in a more favourable position than other Middle Eastern countries because of its larger size, its snowy mountains, and its climate, with its abundant precipitation. Consequently, it is perceived as holding the key to the solution to Middle Eastern water shortages. In this context, several water-related projects involving Turkey have been proposed as solutions to the water shortages of its neighbouring countries. In none of these proposals has Turkey's water potential been realistically assessed. None of these proposals can be ratified, designed, or carried out by Turkey without dependable data and realistic assessments.

Available data on freshwater resources in the region indicate that Turkey (as well as Iraq) does in fact have more water per capita than other Middle Eastern countries, but this is not sufficient to classify Turkey as a water-rich country. In water-related literature, hydrologists use commonly accepted criteria to determine relative water abundance (Falkenmark 1989; Naff 1993; Serageldin 1995). If we assess Turkey's water resources according to accepted, established parameters, it is incorrect to categorize Turkey as a water-rich country. To be rich in water resources, a country must have more than 10 000 m^3/person per year. Water supplies of between 1 000 and 2 000 m^3/person per year make a country water stressed. When the figure drops below 1 000 m^3/person per year, the country is

Table 1. Availability of water by region in the world.

Region	Annual internal renewable water resources		Percentage of population living in countries with stressed and scarce annual per capita water resources	
	Total (1 000 km^3)	Per capita (1 000 m^3)	<1 000 m^3	1 000–2 000 m^3
Sub-Saharan Africa	3.8	7.1	8.0	16.0
East Asia and the Pacific	9.3	5.3	<1.0	6.0
South Asia	4.9	4.2	0.0	0.0
Eastern Europe and former Soviet Union	4.7	11.4	3.0	19.0
Other Europe	2.0	4.6	6.0	15.0
Middle East and North Africa	0.3	1.0	53.0	18.0
Latin America and the Caribbean	10.6	23.9	<1.0	4.0
Canada and the United States	5.4	19.4	0.0	0.0
World	40.9	7.7	4.0	8.0

Source: World Bank (1992).

classified as water scarce, and this usually manifests itself in severe constraints on food production, economic development, and production of natural ecosystems (Table 1).

Turkey's water potential

All of the water resources of Turkey are continuously monitored in a large web of hydrological and meteorological gauge stations throughout the country. Therefore, the data used in this study are accurate and up to date.

The climate within Turkey varies from region to region. Rainy weather during all four seasons is only characteristic of the northern part of Turkey. In the Mediterranean region of the country, the weather is mild and rainy in winter but hot and dry in summer. In the middle, eastern, and southeastern parts of Anatolia, a very large portion of Turkey, the weather is usually drier than in the other regions. Prevailing weather in this region is very hot and dry during summer, and there is less precipitation during the winter as well. In and around Ankara, for instance, a significant water shortage, specifically for agricultural activities, occurs from April until the first week of October (Thontwaite et al. 1958).

Besides seasonal variation, immense differences in precipitation are also found from region to region and from year to year. For example, 63.3 mm of rain fell in Himmetdede, Kayseri (an important province in Middle Anatolia) in 1933, whereas Rize (a province in Northern Anatolia) had 4 043.3 mm of precipitation just 2 years earlier. Urfa, an important province in Southeast Anatolia, only receives an average of 3.9 mm of rain in the summer months (June, July, August), which is the most important period for agricultural cultivation (SHW 1997). Average annual precipitation is 643 mm in Turkey as a whole but changes from region to region and from year to year, ranging from 250 mm in some regions in some years to 3 000 mm in other regions in other years (Altinbilek and Pasin 1998). In short, average annual rainfall varies greatly according to season, year, and region of the country. Therefore, water shortages are an important problem, specifically for agriculture, in Middle Anatolia and in Southeast Anatolia, where an immense water project is currently under way. Irrigation is essential to sustaining and increasing agricultural productivity in these regions. Moreover, many big cities, such as Istanbul and Ankara, experience severe water shortages for domestic and industrial uses during the summer months.

The 643 mm of average annual precipitation in Turkey translates into an average annual water volume of 501 Gm^3 (Figure 1). Of this amount, 186 Gm^3 ends up as surface runoff (Table 2). Some 274 Gm^3, or about 55% of total precipitation, is lost to transpiration and evaporation. Another 69 Gm^3, about 14% of total precipitation, feeds the underground water system. Of this amount, 28 Gm^3 returns to the surface via springs and joins the river systems. In addition, 7 Gm^3 of water comes into Turkey from neighbouring countries. So, altogether (158 + 28 + 7), Turkey's renewable surface-water potential is 193 Gm^3, but the country

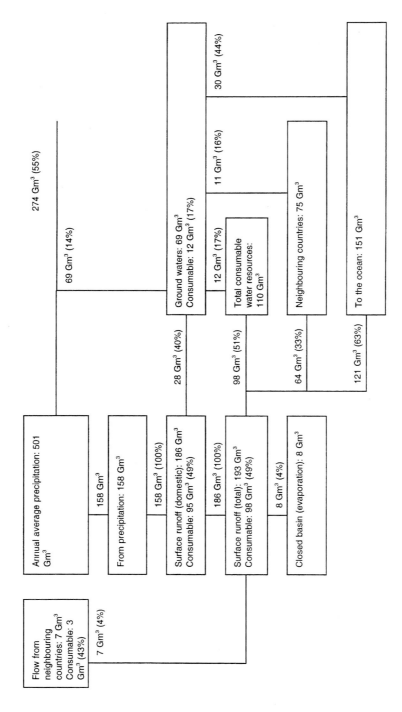

Figure 1. Turkey's water budget.

Table 2. Water potential of Turkey.

Source and origin	Average annual precipitation (mm)	Average annual water volume (Gm³/year)	Flow (Gm³/year)	Economically consumable (Gm³/year)
Surface water in Turkey	643	501	186	95
Surface water outside Turkey			7	3
Underground				12

Source: Altinbilek and Pasin (1998).

cannot use or harness the entire 193 Gm³ because of technological, topographical, and geological constraints. An estimated 95 Gm³/year of Turkey's surface-water runoff cannot be used, but some 98 Gm³ can be. Of this amount, 95 Gm³ originates in the country, whereas 3 Gm³ is transboundary water that originates in neighbouring countries. Some 12 Gm³ of renewable underground water flows into the sea and to neighbouring countries, and this water can be tapped. Therefore, Turkey's total renewable water potential is 205 Gm³ (193 + 12) a year, and of this amount 110 Gm³ (98 +12) can be used economically.

This country of 65 million people has an average annual renewable water potential of 205 Gm³, or about 3 150 m³/person per year, which is far below the 10 000 m³ parameter needed to classify a country as water rich. Taking into consideration the economically usable water potential of the country (110 Gm³), the available annual per capita water goes down to about 1 700 m³, which would make Turkey a water-stressed country. Furthermore, rapid population growth, industrialization, and rising standards of living will decrease the annual per capita renewable water potential to 2 500 m³ by 2000 and to 2 000 m³ by 2010. If we estimate the economically usable per capita annual water potential, we can project a severe situation in which available water goes down to 1 580 m³/person per year, or even less, by 2000. As can be seen from these data, Turkey's water resources are far from abundant. Table 2 shows that it has only about one-fifth or one-sixth of the water available in water-rich regions, such as the Caribbean, Latin America, North America, and even Western Europe.

There are 26 hydrologic basins in Turkey (Table 3). Of these, 22 are river basins and the other four are enclosed basins that have no flow to the sea. Two river basins (the Euphrates and the Tigris) contain the largest volume of flow of all the rivers in Turkey, 28.5% of the nation's total surface flow (17% in the Euphrates and 11.5% in the Tigris). Dogu Karadeniz (East Black Sea), with an 8% contribution, Dogu Akdeniz (East Mediterranean), with a 6% contribution, and Antalya, with a 5.9% contribution, are other relatively water-rich basins.

Turkey has built hundreds of dams and hydroelectric power plants, and it has carried out other water-related projects to harness water, produce energy, and irrigate arid lands, but this still does not mean that Turkey has fully benefited from these resources. About 37 Gm^3 of Turkey's 110 Gm^3 of usable water is actually used. Almost 33% of economically usable water can actually be used at present. The remaining 67% of economically usable water, which Turkey desperately needs for economic development, still flows freely into the sea.

In 1997, 681 dams higher than 15 m were already built or under construction to harness the economically usable surface water of Turkey. Of these dams, 465 are now in operation and harness about 30% of this water. The remaining 216 dams are still under construction. Moreover, many projected and planned dams will be used to harness the remaining 67% of economically usable water in Turkey to meet future needs.

Some of the main purposes of these dams are recreation, flood protection, domestic water supply, irrigation, and energy production. The majority of Turkish dams are for domestic water supply and irrigation, and many of them are multi-purpose dams. These dams generate electrical energy and also supply water for irrigation or domestic needs. Of the 681 dams, 63 are used only to generate electrical energy. Multipurpose dams that generate energy account for almost 10% of all dams. This low figure for hydroelectric generation is a result of the priority given to developing water resources mainly for domestic and agricultural use.

Despite its own growing need for water, Turkey is still willing to export some of its water to neighbouring countries to relieve their shortages. The main water resources that can be used for this purpose are in the southern basins of Turkey. These basins (Eastern Mediterranean, Antalya, Western Mediterranean, Seyhan, and Ceyhan) constitute almost 25% of Turkey's total renewable water potential. Several dams are in operation on these rivers, and several more are under construction, but much of the water of these rivers still flows into the Mediterranean Sea, without being used. This water could be used to alleviate water shortages in some countries of the Middle East, as well as in the parts of Turkey,

Table 3. Turkey's annual average water potential by basin.

Basin	Average annual flow (Gm³)	Contribution to total (%)
Firat (Euphrates)	31.61	17.0
Dicle (Tigris)	21.33	11.5
Dogu Karadeniz	14.90	8.0
Dogu Akdeniz	11.07	6.0
Antalya	11.06	5.9
Bati Karadeniz	9.93	5.3
Bati Akdeniz	8.93	4.8
Marmara	8.33	4.5
Seyhan	8.01	4.3
Ceyhan	7.18	3.9
Kizilirmak	6.48	3.5
Sakarya	6.40	3.4
Coruh	6.30	3.4
Yesilirmak	5.80	3.1
Susurluk	5.43	2.9
Aras	4.63	2.5
Konya	4.52	2.4
Buyuk Menderes	3.03	1.6
Vangolu	2.39	1.3
Kuzey Ege	2.09	1.1
Gediz	1.95	1.1
Meric	1.33	0.7
Kucuk Menderes	1.19	0.6
Asi	1.17	0.6
Burdur Goller	0.50	0.3
Akarcay	0.49	0.3

Source: SHW (1998).

that experience water shortages. Several projects have been devised to use the water of Turkey's Mediterranean rivers for this purpose.

One of the best known projects to this effect is the Manavgat Water Supply Project. This project and others have been devised to alleviate water shortages in some parts of Cyprus, the Middle East, and Turkey. In the last week of July 1998, a project went into action to transport water from Turkey's Mediterranean rivers to the Turkish Republic of Northern Cyprus in big balloons, with the Manavgat facility as the point of loading.

Southeast Anatolia Project

A foremost aim of Turkey is to eliminate interregional economic and social imbalances within its borders. The optimal use of land and water resources is an important means to achieve this goal. The most important investment scheme this country has undertaken in this endeavour is the Guneydogu Anadolu Projesi (GAP, Southeast Anatolia Development Project).

Goals of GAP

The Turkish government has designed and implemented this large project in Southeast Anatolia for two main reasons. First, Southeast Anatolia is endowed with good water and land resources, and Turkey wants to use these resources optimally for the sake of the entire region, as well as for Turkey as a whole. GAP is being developed on the Euphrates and Tigris rivers and their branches that originate in Turkey. These watercourses supply the majority of Turkey's total surface water, flowing later through Iraq and Syria to reach the Persian or Arabian gulf. Second, Southeast Anatolia is the most backward region of the country. There are big economic and social disparities between this region and the rest of Turkey. For instance, per capita income in the region is 47% lower than the per capita income of Turkey as a whole (GAP RDA 1995). In other words, the average per capita income of Turkey is more than twice that of Southeast Anatolia. Moreover, many economic and social indicators, such as per capita electrical energy consumption, number of hospital beds per 10 000 people, and manufacturing's share of the gross national product in the region, make it clear that it desperately needs investment. Development of this region is key to eliminating economic disparities between Southeast Anatolia and other parts of Turkey. The project on the two transboundary rivers aims at eradicating regional inequality and promoting economic growth and social stability in this region.

Initial work on the Euphrates River was started by the Euphrates Planning Authority, established in Diyarbakir in 1961. Coming out of this work, the *Reconnaissance Report for the Euphrates Basin* appeared in 1964, and it clarified the irrigation and energy potential of the basin concerned. Further studies have subsequently been carried out and published. Meanwhile, work of a similar nature has been carried out on the Tigris Basin by the Diyarbakir Regional Directorate of the State Hydraulic Works (SHW 1998). These studies made clear that the Euphrates and Tigris rivers have significant potential to help the region develop economically. Finally, in 1977, projects related to these two basins were merged and adopted as an integrated, multisectoral single project, under the title of the GAP.

The GAP area lies in southeastern Turkey and takes in nine provinces. This region is part of Upper Mesopotamia, which was the cradle of the ancient Mesopotamian civilization. The total area of the project is about 10% of Turkey, and according to recent statistics, includes about 9.5% of Turkey's total population. The project envisages the construction of 22 dams, 19 hydroelectric power plants, and 2 irrigation tunnels on the Euphrates and Tigris rivers and their tributaries. The major element of the project, the Ataturk Dam and the Sanliurfa Tunnel System, are already completed and in operation.

When the whole project is completed, 1.7 million ha of land will be irrigated, the ratio of irrigated land to the total GAP area will increase from 2.9% to 22.8%, and the area of rain-fed agriculture will decrease from 34.3% to 7%. In addition, 27 TWh of electricity will be generated annually from an established capacity of 7 460 MW. The area to be irrigated is 19% of all economically irrigable land in Turkey (8.5 million ha), and annual electricity generation will come to 22% of the country's economically feasible hydroelectric power potential (that is, 118 TW).

The economic benefits expected from the project are substantial. Many agricultural crops will double or even triple. GAP will provide Turkey with food self-sufficiency and will create 3.3 million jobs. Turkey's national income will be 12% higher than it would otherwise have been, and the gross regional product of Southeast Anatolia will increase by more than fourfold. Urbanization will receive a boost in the region (actually it has already been boosted), and rural migration will slow down considerably.

The objectives and the main features of the integrated project are outlined in the GAP Master Plan, completed in 1989 (GAP 1989). The Prime Minister's

Office published the strategy adopted in the GAP Master Plan, with the following four basic components:

- To efficiently develop and manage soil and water resources for irrigation, industrial, and urban uses;

- To improve land use through optimal cropping patterns and better agricultural management;

- To promote manufacturing with emphasis on agro-related industries and those based on indigenous resources; and

- To provide better social services, education, and employment opportunities in order to control migration and attract qualified personnel to the area.

In short, the GAP Master Plan's basic development scenario is to transform the region into an export base for agroindustrial products. The project was initially conceived only to irrigate arid lands in the region and generate hydroelectric energy from the Euphrates and Tigris. However, the objectives of the project have been expanded to include overall socioeconomic development, and GAP is now a multifaceted development project that will bring economic, social, and cultural changes affecting not only the local region but also the country as a whole. What started out as a simple project for hydroelectric power plants and irrigation systems has turned into a massive project with interests in urban, rural, and agricultural infrastructure, transportation, industry, education, health, housing, and tourism, and investments in many other fields.

Social aspects of GAP

Until a few years ago, the main emphasis of GAP was on planning, construction, start-up, and operation of physical components, such as dams, hydroelectric plants, and irrigation systems. The Turkish government has made immense efforts to implement the largest and most comprehensive regional development investment plan in Turkey during its Republican era. In financial terms, the project had, in 1998, a realization rate of more than 40%. The rates of realization for the sectors of energy and agriculture were, respectively, 73% and 11%. Having already accomplished many of its goals, the project has reached a new phase, and policymakers have adopted a new approach. The main features of this new approach are sustainability and human development, no longer just physical implementation. As a result, the social aspect of the project, along with water and land-resource development, has become one of the main concerns of the GAP administration.

As defined by the GAP Administration, the objective of "sustainable human development" was "to take economic growth into the human development perspective and to convert the social transformation, which will cover the whole region, into participatory solutions of an ecological, cultural and local nature" (GAP RDA 1995). A symposium on Sustainable Development and GAP was held by the GAP Administration and the United Nations Development Programme in March of 1995. Based on the results of this seminar and the objectives and targets of the GAP Master Plan, the following sustainability goals have been adopted for the development process:

- To increase investment to the highest possible level in order to accelerate the improvement of economic conditions in the region;

- To enhance health care and educational services so that they reach national standards;

- To create new employment opportunities;

- To improve the quality of life in the cities and build urban and social infrastructure so as to create healthier urban environments;

- To complete the rural infrastructure for optimal irrigation development;

- To increase inter- and intraregional accessibility;

- To meet the infrastructure needs of existing and new industry;

- To protect water, soil, air, and associated ecosystems as a priority consideration; and

- To enhance community participation in decision-making and project implementation.

To give a higher priority to the social aspect of the project, the GAP Administration planned and carried out community-survey studies to make the people of the region fully aware of the nature of GAP and raise interest in becoming an integrated part of it. The survey studies were

- Trends of Social Change in the GAP Region;

- Population Movement in the GAP Region;

- The Status of Women in the GAP Region and the Integration of Women into the Process of Development and Resettlement in Areas Which Will Be Affected by Dam Lakes;

- Management, Operation and Maintenance of GAP Irrigation Systems; and

- Socioeconomic Aspects.

Based on the results of these survey studies, the Gap Administration prepared a document called the *GAP Social Action Plan*, in which the human aspects of development are emphasized. This plan, with a sustainable, participatory and integrated approach, constitutes the framework for the implementation of mid-term phases of the project. The objectives of the GAP Social Action Plan are as follows:

- To underline the human factor in GAP and relate this basic element to each project developed in accordance with the GAP Master Plan;

- To ensure the integration of different groups and layers of society in GAP Region into the development process;

- To enhance the efficiency and coverage of social services in the region in a way that eliminates disparities between this region and other regions of the country;

- To bring about sustainable development by ensuring people's participation in the design and implementation of projects; and

- To produce strategies and policy proposals to guide planners and implementors.

In line with the goals of the development process and objectives of the Social Action Plan, some pilot projects have been implemented in the region and then expanded. Multi-purpose Community Centres (MPCCs) and GAP Entrepreneur Support and Guidance Centres (GAP ESGCs) are worth mentioning.

MPCCs are centres where training is provided for women and girls in literacy, health care, maternal care, child care, nutrition, home economics, and income-generating handicrafts. A participatory and integrated approach is the basic policy of all MPCCs, and more than 10 of them have been established in the region. They have already become one of the main instruments of the government to improve the status of women and children and the living standards of all people in the region.

The purpose of the GAP ESGCs is to encourage private-sector investment in the GAP provinces and to provide consulting services to entrepreneurs before and after they make their investments. GAP ESGCs are an important tool to accelerate economic development in the region.

The environment is another main concern of the GAP Administration, and several preliminary environmental studies have been carried out. The main objective of these studies is to identify existing and possible future environmental problems that could be caused by implementation of the irrigation projects, dams, and hydroelectric power plants and to make recommendations to limit environmental damage without interfering with development objectives.

Environmental policies have also become one of the main concerns in relation to sustainable economic and human development. In this context, the Ministry of the Environment and the GAP Administration signed a joint protocol laying down principles of cooperation for the two organizations to identify environmental

Table 4. Foreign finance in the GAP region.

Source of Credit	Amount of credit (million USD)
US Exim Bank	111
Swiss Commercial	467
Swiss-German Commercial	782
European Investment Bank	104
World Bank	120
European Council Social Development Fund	183
Italian Government	85
French Government	33
German Government	15
Austrian Government	200

Source: GAP RDA (1995, 1998).
Note: USD, United States dollar.

problems in the region and the relevant measures to address these problems. The protocol was signed on 21 April 1998.

Current status of gap development

The total estimated cost of the GAP development is 32 billion United States dollars (USD). As of the end of 1997, total spending on the project had reached 12.6 billion USD, at a financial realization rate of 41.3%. GAP is largely financed by national resources, that is, the budget of the Turkish government. Nevertheless, a combination of foreign suppliers' credits, loans from international agencies and foreign banks, and state export-insurance schemes are used to finance the various GAP component projects: dams and hydroelectric power plants, water infrastructure, health projects, agricultural research, and new and modern irrigation systems, among others. Table 4 shows that about 2.1 billion USD worth of external credit, secured from various sources, has contributed to GAP development.

Because of Turkey's economic problems in the period of 1990 to 1998, its share of GAP investment allocations in the Annual Investment Programs declined from 8.1% to 6.6%, at 1998 fixed prices. This created a bottleneck from the point of view of implementing the project within the projected time frame. To alleviate this bottleneck, the Turkish Government decided to create new financial sources for GAP projects, to add to the already existing national and international sources. As a result of this decision, some new financial mechanisms, such as Build–Operate–Transfer (BOT), were created to finance some GAP projects. For instance, construction on the Birecik Dam and Hydroelectric Power Plant is being carried out on the BOT basis.

Completed GAP projects, such as the Ataturk and Karakaya dams, are generating a substantial amount of hydroelectric energy since they went into operation. There is a significant change in crop patterns and a large increase in agricultural incomes from newly irrigated lands. In other words, some of the GAP investments are already starting to pay dividends. This situation has created new motivation for the Turkish government to generate new financial sources. If the project receives a new financial boost, it could be completed earlier than originally planned. The Ataturk and Karakaya dams, the most important investments of GAP, have generated almost 135 TWh of electrical energy, as of 15 June 1998, for a monetary value of 8 billion USD. If we were to compare this amount to alternative sources of energy, it would correspond to having to import 33 million t of fuel oil or 25.5 Gm^3 of natural gas.

In the Euphrates and Tigris basins, the area brought under irrigation for the 1998 irrigation season reached 174 080 ha, almost 10% of the projected irrigation area within the scope of GAP (1.7 million ha). Irrigation for another 11% is now under construction (183 995 ha).

There are striking changes in crop patterns in the region, now that there is irrigation. Before, wheat, barley, and lentils used to be the main crops. Now, cotton, maize, peanuts, sunflowers, soybeans, and vegetables are being produced, and they contribute to the growth of the agricultural industry. The biggest change is in the amount of land used for cotton. As of the end of 1997, about one-third of the cotton harvest in Turkey was carried out in the GAP region. Cotton is grown on 38 664 ha, part of a total of 60 000 ha of land thus far brought under irrigation in the Sanliurfa-Harran Plain.[1] The total value of agricultural production in the

---

[1] A population of 66 360 in 104 villages located in 60 000 ha of land brought under irrigation.

Table 5. Economic returns from 60 000 ha of land opened to irrigation in the Sanliurfa-Harran Plain.

Indicator	Prior to irrigation (USD)	After irrigation (USD)
Agricultural Income	31.5 million	120.5 million
Agricultural value added	600/ha	1 619/ha

Source: GAP RDA (1998).
Note: USD, United States dollars.

region is estimated at 120.5 million USD, up from 31.5 million USD (Table 5). These figures, which refer to a small portion of the total area to be irrigated, give a general idea of the economic returns to be reaped when the project is fully completed.

Irrigation and the resulting increase in agricultural production have already resulted in positive developments in terms of industrial entrepreneurship in the region, and, as mentioned previously, GAP ESGCs have had an important function in this development. Several Organized Industrial Districts (OIDs) and Small Industrial Estates (SIEs) have been established and are being expanded in the region to foster this development by providing settlements and infrastructure for small and medium-sized enterprises. As of the end of 1997, there were 3 OIDs covering a total of 1 060 ha. The 1998 Annual Investment Program includes 11 new OIDs and three water-treatment projects in the GAP region. Some 18 SIEs were active in the region in 1998.

Despite these developments, the project is still far from its targets. As mentioned above, there are big problems in financing. If the government does not want to revise the timetable for project implementation, it must generate new sources of financing and invest more money in the project. There seems to be a desire for this in the Turkish government. On the down side of investment, accelerated economic development and the resulting increase in demand for land have caused real-estate prices to skyrocket. Several financially powerful companies and individuals started to buy up real estate in the region to sell later for large profits. A natural result of this is that much of the land is now in the hands of a few people. In other words, a new monopoly of land ownership has become an economic problem in the region.

As can be seen, the impact of GAP on the region's as well as Turkey's economic, social, and cultural life is enormous. Water in the Euphrates and Tigris rivers has already started to improve the standard of living of local citizens by increasing income levels, providing employment, and bringing stability to the region. Using the water of the Euphrates and Tigris rivers has become one of the prerequisites for the Turkish government to make the region economically prosperous and socially and politically stable. The contribution of the production of hydroelectric energy to the Turkish economy and economic returns on irrigation being reaped by the people of the region have spurred the Turkish government on to want to complete the project as soon as possible.

Turkey, as an oil-poor, developing country, needs to use its water for the economic and social development of Southeast Anatolia, as well as for that of the country as a whole. At the same time, Turkey should be very diligent when it uses the waters of the Euphrates and Tigris rivers so as to prevent any adverse effects on neighbouring countries or the environment. By adopting a sustainable approach to economic development, using improved irrigation techniques to conserve water, and releasing more water than the amounts agreed on from Turkish territory, Turkey has constantly revealed its good will toward its southern neighbours. When neighbouring states choose to adopt a rational attitude toward GAP, the benefits of this project will not be limited to Southeast Anatolia, nor solely to Turkey, but will also produce far-reaching positive effects for the whole Eastern Mediterranean region.

References

Altinbilek, D.; Pasin, S. 1998. Hydroelectric energy potential of Turkey and current situation. State Hydraulic Works, Ankara, Turkey.

Falkenmark, M. 1989. The massive water scarcity now threatening Africa — why isn't it being addressed? Ambio, 18(2).

GAP (Guneydogu Anadolu Projesi [Southeast Anatolia Project]). 1989. Southeastern Anatolia Project Master Plan study. Nippon Koei Co. Ltd; Yuksel Proje A.S. [joint venture], Ankara, Turkey.

GAP RDA (GAP Regional Development Administration). 1995. South-eastern Anatolia Project: foreign resource use. Prime Minister's Office. Ankara, Turkey.

———— 1998. South-eastern Anatolia Project: latest state as of April 1998. Prime Minister's Office; Afsaroglu Publications, Ankara, Turkey.

Naff, T. 1993. Water: that peculiar substance. Research and Exploration, 9 (special issue), 6–18.

Serageldin, I. 1995. Toward sustainable management of water resources. World Bank, Washington, DC, USA.

SHW (State Hydraulic Works). 1997. Statistical bulletin with maps, 1997. Ankara, Turkey.

———— 1998. Turkey's Hydro-electric energy potential and current situation. Ankara, Turkey.

Thontwaite, C.W.; Mather, J.R.; Carter, D.B. 1958. Three water balance maps of Southwest Asia. Laboratory of Climatology, Centertown, NJ, USA.

World Bank. 1992. World development report 1992: development and the environment. Oxford University Press, Oxford, UK.

Chapter 7

Transporting Water by Tanker from Turkey to North Cyprus: Costs and Pricing Policies

Hasan Ali Biçak and Glenn Jenkins

Introduction

North Cyprus is in a semi-arid region where average annual rainfall varies from 200 to 600 mm. From the beginning of the century, it has experienced a reduction in average annual rainfall: from 440–450 mm at the beginning of the century, to 402 mm from 1941 to 1972, to 382.4 mm from 1975 to 1993 (Biyikoglu 1995). In addition, overextraction of water from aquifers has resulted in seawater intrusion all over the island. Seawater intrusion in Gazimagusa and Gecitkale aquifers has been so severe that the water is no longer potable, and water stations had to be set up to sell fresh water. Also, because of water shortages and the use of saline water for irrigation, a large number of citrus trees have died, and the land they grew on is no longer irrigated. Between 1976 and 1996, land used for citrus production fell from 74 710 donums (1 donum [dn] of land is equal to 0.1338 ha) to 47 700 dn. In the same period, total irrigated land fell from 116 400 dn to 74 044 dn (MOAF 1997a, b).

The land area of North Cyprus is 2 465 552 dn, of which 1 392 123 dn (57%) is agricultural. About 881 481 dn (63.05%) of this land is cultivated, of which 805 437 dn (91.6%) is rain fed and 74 044 dn (8.4%) is under irrigation (MOAF 1997b).

NB: The authors would like to thank Ali Özdemirag for his extensive assistance in completing the quantitative aspects of this study.

Previous studies, which did not take account of the sharp fall in irrigated area, estimated demand for water in North Cyprus at between 190 and 197 Mm^3 and the actual supply of water at between 110 and 125 Mm^3, without explaining how, in practice, the deficit was made up (Numan and Agiralioglu 1995; TCW 1996[1]). If one does take into account the sharp fall in irrigated land, distinguishing between land irrigated with traditional and that irrigated with modern methods, demand for water in North Cyprus can be estimated at 106.6 Mm^3 for 1996 (Biçak and Özdemirag 1997). In the study by Biçak and Özdemirag, water demand was estimated as 87.5 Mm^3 (82.1%) for agricultural use, 17.1 Mm^3 (16.1%) for household consumption (including the armed forces, seasonal workers from Turkey, students and the tourists), 1.3 Mm^3 (1.2%) for animals, and 0.7 Mm^3 (0.7%) for commercial and industrial use, giving a total demand of 106 Mm^3.

As for the potential supply of water, no reliable figure will be available until research currently conducted by the Mines Investigation and Search Institute of Turkey is completed. Previous data show that about 74.1 Mm^3/year can be extracted from the aquifers without depleting them, but it is estimated that overextraction of water from the aquifers could be as high as 28.9 Mm^3/year, giving a total yearly extraction of 103 Mm^3. Guzelyurt aquifer is the biggest on the island, with 37 Mm^3 of safe-yield capacity, and it is believed that 20 Mm^3/year is overextracted from this aquifer alone.[2] An alternative would be to consider rivers and existing dams as providing annually about 13 Mm^3 and 7 Mm^3 of water, respectively, depending on rainfall. On this basis, total potential water supply of North Cyprus is 94.1 Mm^3 (74.1 + 13 + 7). This makes the water deficit about 12.5 Mm^3 (106.6 − 94.1), and the deficit is now being filled by overextraction from aquifers.

Water shortage for domestic and agricultural use in North Cyprus is evident. Various measures are planned, and others have been implemented to increase the supply of water and use it more efficiently. Projects have been proposed to import water from Turkey by tanker, in large water bags or by pipeline. The water-bag option has been tried, starting from 25 July 1998 (Kibris 1998a). It is estimated that water bags with a 10 000-m^3 capacity can bring 3 Mm^3 of water in

---

[1] TCW (Technical Committee on Water). 1996. Summary report on the studies carried out by the Committee. TCW, Nicosia, North Cyprus. Unpublished report, 31 Jan.

[2] State Planning Organization of the Turkish Republic of North Cyprus, Prime Ministry, Nicosia, North Cyprus, 1992.

1 year. The water comes from the Soguksu River in Anamur, Turkey. An increase in the capacity of the water bags to 30 000 m^3 would enable 7 Mm^3 of water to be imported annually. This is the maximum amount that the system in North Cyprus can allow to be pumped. It is pumped from Kumkoy to Serhatkoy, and then on to Dikmen (where the main reservoirs are situated), and from there to Nicosia and Gazimagusa.

Another important project (currently implemented in the Guzelyurt area) aims to prevent the excessive use of water by converting traditional irrigation systems to modern ones. A large amount of water will be conserved, salination will be prevented, and the productivity and quality of agricultural output will improve. Production costs will decrease because less fertilizer will be needed. Currently, of all irrigated land (74 044 dn), some 66 084 dn (89.2%) is irrigated by traditional methods, and 7 960 dn (10.8%) is irrigated by modern methods, with sprinklers (2 989 dn, or 37.6%) and drip irrigation (4 971 dn, or 62.4%) (MOAF 1997b). The project started by converting 10 000 dn from traditional to modern irrigation practices on citrus farms in the Guzelyurt area. Authorities are planning to convert 10 000 dn of citrus land every year to modern irrigation practices, so that all crops will be irrigated this way by 2001 (Sevki 1997). Considering that 1 dn of citrus land uses 1 420 m^3 of water annually with traditional methods, but only 710 m^3 with modern methods, this is expected to save a large amount of water, potentially as much as 46.9 Mm^3 (710 × 66 084), if the project is implemented successfully.

To ease the water shortage of North Cyprus, there is a proposal to import water by pipeline from Anamur or Manavgat, Turkey. If this project is implemented, 75 Mm^3 of water could be brought to Kumkoy, North Cyprus, from where it would be further distributed. The Council of Ministers of Turkey has decided to implement the project through the Turkish firm, ALARKO Holding Company. This project appears to be financially infeasible for Turkey if the project's revenues are limited only to those gained from selling water in North Cyprus. It might become financially feasible if more water is sold to South Cyprus or to other Middle Eastern countries (Biçak 1996). It might be worthwhile, in political terms, for Turkey to build a permanent pipeline supply link to North Cyprus. Once the capital investment is made, the marginal pumping costs and operating expenses would be very low.

Another project to supply water to Nicosia and Gazimagusa is to build a dam in the Yesilirmak area, where, depending on precipitation, about 8–12 Mm^3

of water flows underground into the sea (Özdemirag 1998). This project is now on hold because of a conflict between the views of the government and those of local villagers affected by the project.

Rehabilitation of the Haspolat Sewage Treatment Plant is expected to be completed by the end of 1998. Once it is completed, it will provide 3.5 Mm^3 of water for agricultural use. Although the plant has been operating since 1980, it provides a very limited amount of water for agricultural use because much of the proposed infrastructure is not in place (Oznel et al. 1997; Kibris 1998b).

Setting aside all these projects either proposed, planned, under construction, partially implemented, or fully completed, this chapter will now turn to its main purpose, a feasibility study of importing water by tanker from Turkey to North Cyprus. The model chosen enables us to separate the effects of various components likely to have an impact on the unit cost of water and the financial outcome of the project. These components include inflation, billing cycle, payment terms, and a system for adjusting tariffs for inflation. A sensitivity analysis will enable us to identify the key variables that may affect the outcome of the project.

Methodology

To analyze the financial feasibility of importing water from Turkey to North Cyprus by tanker, a proforma cash-flow statement was constructed. Cash flow considers all revenues and expenditures throughout the life of the project. Net cash flow is projected from two points of view: equity (the owner) and total investment (the banker). Net cash flow projected from the equity point of view includes loans and repayment of the principal and interest, whereas net cash flow projected from the total-investment point of view excludes these items. The total-investment point of view analyzes the strength of the project in the absence of such financing arrangements (Harberger and Jenkins 1998). In this study, net present value (NPV) is calculated from the point of view of the owner (equity), using a real (inflation-free) discount rate of 12%.

North Cyprus, not having a currency of its own, uses the Turkish lira (TRL) as its medium of exchange and, along with Turkey, experiences annual inflation rates of about 80%. In such an inflationary environment, the length of the billing and payment cycle, as well as the system for adjusting water tariffs for inflation would play a significant role in the financial performance of the project.

The impact of inflation on the project is incorporated into the model by constructing net cash flows in nominal prices first (using assumed rates of nominal price adjustment). These nominal values are then converted into real prices by deflating them with a general price index that reflects the assumed overall rate of inflation in North Cyprus (Harberger and Jenkins 1998).

When net cash flow is calculated in this way, we can estimate the transportation cost per cubic metre of water from Turkey to North Cyprus, excluding all infrastructure investments and operating costs in North Cyprus itself. The cost per unit of delivered water, evaluated at its point of entry in Kumkoy, excludes the cost of leakage in the system, the financial effects of time lags in billing and payment for the water, and administrative lags in adjusting nominal tariffs for inflation. Next, we estimate the unit cost of the water, including the cost of water leakage in the distribution system, and, finally, the last set of cost calculations introduces alternative scenarios or combinations of administrative ways of handling accounts receivable, accounts payable, and lags in adjusting the nominal prices of water for inflation. These calculations will show the financial implications of various alternative pricing policies as used by municipalities in North Cyprus.

Project description

Objective and scope of the project

The objective of the project is to import fresh water from Turkey by tanker to meet the demand for potable water in households. The project does not aim to provide water for agricultural use or for recharging the aquifers badly depleted or affected by the seawater intrusion.

Currently, fresh water is pumped from Kumkoy to Serhatkoy and from there to the main reservoir in Dikmen. Then the water is distributed to Nicosia and Gazimagusa. Kumkoy is supported by 14 wells and sends 9 000 m³/day of water to Serhatkoy. Adding 3 500 m³ of water from four nearby wells, Serhatkoy pumps a total of 12 500 m³ /day. Of this amount, 3 500 m³ is sent to the Turkish part of Nicosia through South Cyprus, and 9 000 m³ is sent to Dikmen, from which point 3 000 m³ is sent to Gazimagusa and 6 000 m³ is sent to Nicosia. This amount of water is insufficient, and the quality of the water is very poor. The pipes have a

Table 1. Distribution of water and existing infrastructure.

Capacity of existing infrastructure	
From Kumkoy and Serhatkoy to Dikmen (m³/hour)	750
From Kumkoy and Serhatkoy to Dikmen (m³/day)	18 000
From Kumkoy and Serhatkoy to Dikmen (m³/year)	6 570 000
Sources and distribution of water	375
From 14 wells to Kumkoy (m³/hour)	
From 14 wells to Kumkoy (m³/day)	9 000
From Kumkoy to Serhatkoy (m³/hour)	375
From 4 nearby wells to Serhatkoy (m³/hour)	145
Total amount of water arriving in Serhatkoy (m³/day)	12 500
From Serhatkoy through South Cyprus to Nicosia (m³/day)	3 500
From Serhatkoy to Dikmen (m³/day)	9 000
Total amount of water distributed from Serhatkoy (m³/day)	12 500

Source: Data on the tanker project were obtained from the Undersecretary's and the Port General Directorate's offices of the Transportation Ministry, and data on the pipeline project were obtained from the Water Works Office of the Ministry of the Interior of the Turkish Republic of Northern Cyprus.

diameter of 18 inches (1 inch = 2.540 cm), and they cannot transport more than 18 000 m³/day, or 6.57 Mm³/year (Table 1).

Manavgat, on the south coast of Turkey, was chosen from a number of possible sources for water to ship to North Cyprus because it already had the necessary infrastructure on land, and some of the sea structures were expected to be completed shortly. Currently, at Manavgat, 500 Mm³ of fresh water flows annually into the sea. Once the land and sea infrastructure is completed, it would be possible to export water to other Mediterranean countries. Manavgat, Turkey, and Kumkoy, North Cyprus, are 248 km apart. Considering the volume that the distribution system in North Cyprus can handle (6.57 Mm³/year), one tanker with 40 000-m³ capacity, making 175 trips a year, could transport 7 Mm³ of water. The tanker is assumed to operate 320 days a year, staying nonoperational 45 days a year for maintenance and repairs and for days when weather conditions are unsuitable for navigation (Table 2).

Table 2. Capacity of tanker and volume of water to be imported.

Number of nonoperational days per year	45
Number of operational days per year	320
Distance between Manavgat, Turkey, and Kumkoy, North Cyprus (km)	248
Tanker's average speed (km/hour)	20.8
Time to travel one way (hours)	12
Time for loading in Manavgat (10 000 m³/hour)	5
Time for connecting, disconnecting, and formalities in Manavgat (hours)	3
Time for discharging in Kumkoy (hours at 4 000 m³/hour)	10
Time for connection, disconnection, and formalities in Kumkoy (hours)	2
Total time for one round trip (hours)	44
Total number of trips per year	175
Total volume of water per trip (m³)	40 000
Total volume of water per year (m³)	7 000 000

Source: Data on the tanker project were obtained from the Undersecretary's and the Port General Directorate's offices of the Transportation Ministry, and data on the pipeline project were obtained from the Water Works Office of the Ministry of the Interior of the Turkish Republic of Northern Cyprus.

Total investment and operating costs

For this analysis, it is assumed that the tanker will be owned and operated under normal private financial arrangements. Although the installations in Manavgat, Turkey, are near completion, the government of North Cyprus needs to start building the necessary facilities on land and offshore in Kumkoy. A port is not required for the tanker in North Cyprus; rather, an offshore mooring system is sufficient. The water will be pumped through a sea-to-land pipeline to the reservoirs at Kumkoy. The existing system at Kumkoy will pump the water to Serhatkoy. To handle the increased capacity of water sent from Kumkoy, the pumping system in Serhatkoy will need to be augmented with two additional pumps. Table 3 shows that the

Table 3. Total investment costs (1998 prices) (USD).

Tanker	
Cost of tanker	8 000 000
Cost of offshore mooring system	2 000 000
Cost of boat for anchoring tanker and connecting–disconnecting pipes	250 000
Offshore pipeline (mooring station to shore, 1.5 km)	
Cost per km	750 000
Cost of offshore pipeline	1 125 000
Land pipeline (shore to Kumkoy: 2 km)	
Cost per km	500 000
Cost of land pipeline	1 000 000
Reservoir at Kumkoy (capacity, 2 x 20 000 m^3)	
Cost per Mm^3	90
Cost of reservoir (90 x 40 000)	3 600 000
New pumps at Serhatkoy (number, 2; capacity, 750 m^3/hour, or 18 000 m^3/day; average horsepower, 375)	
Cost of new pumps (2 x 500 000)	1 000 000
Total investment costs	16 725 000

Source: Data on the tanker project were obtained from the Undersecretary's and the Port General Directorate's offices of the Transportation Ministry, and data on the pipeline project were obtained from the Water Works Office of the Ministry of the Interior of the Turkish Republic of Northern Cyprus.
Note: USD, United States dollar.

total investment cost of the project, including infrastructure and the tanker, will be 16.725 million United States dollars (USD).[3]

Operating costs of the project include crew salaries, salaries for additional employees at Kumkoy, fuel and diesel-oil consumption, and maintenance. Annual total for the salaries of the crew is expected to be 493 200 USD; and for workers at Kumkoy, 76 800 USD. Costs of fuel and oil consumption will be 1 304 926 USD annually. Maintenance costs are expected to be around 147 250 USD. Table 4 gives an itemized breakdown of total annual operating costs (with an additional

---

[3] Data on the tanker project were obtained from the Undersecretary's and the Port General Directorate's offices of the Transportation Ministry, and data on the pipeline project were obtained from the Water Works Office of the Interior Ministry of the Turkish Republic of Northern Cyprus.

Table 4. Total operating costs (USD).

Monthly crew salaries on tanker	
Four captains at 2 000	8 000
Four engineers at 1 800	7 200
One communications officer at 1 300	1 300
Eight above-deck and eight below-deck crew members at 1 200	19 200
Two cooks and four stewards at 900	5 400
Total monthly crew salaries on tanker	41 100
Total annual crew salaries on tanker	493 200
Monthly personnel salaries at Kumkoy	
One captain at 800	800
One mechanical engineer at 800	800
Two boat crew members at 600	1 200
Six Water Resources Department employees at 600	3 600
Total monthly personnel salaries at Kumkoy	6 400
Total annual personnel salaries at Kumkoy	76 800
Fuel oil consumption — 35 t per round trip at 150/t	5 250
Total annual cost of fuel oil	919 579
Diesel oil consumption: 10 tons per round trip at 220/t	2 200
Total annual cost of diesel oil	385 347
Port handling costs at Manavgat at 5 000 per trip	875 789
Annual insurance costs at 2% of the tanker's initial price	160 000
Water cost — 7 million m^3 at 0.15/m^3	1 050 000
Cost of maintenance at 1% of initial price	
Tanker	80 000
Offshore pipeline	11 250
Land pipeline	10 000
Reservoir	36 000
Pumps	10 000
Total annual cost of maintenance	147 250
Miscellaneous (1% of operating costs)	30 580
Total annual operating costs	4 138 545

Source: Data on the tanker project were obtained from the Undersecretary's and the Port General Directorate's offices of the Transportation Ministry, and data on the pipeline project were obtained from the Water Works Office of the Ministry of the Interior of the Turkish Republic of Northern Cyprus.
Note: USD, United States dollar.

1% to cover miscellaneous items), for a grand total of 4 138 545 USD (1998 prices).

Sources of financing

Plans are that 70% of the total investment costs (11.707 5 million USD) would be borrowed in US dollars, directly from Turkey or else from international financial institutions with guarantees from Turkey. The rest of the investment costs (5.017 5 million USD) will be equity financed. The real interest rate on the loan (before risk adjustment) is assumed to be 4%. In addition, there will be a 5% risk premium associated with Turkey. Therefore, the loan would be taken out at a 9% real basic interest rate. Taking into account an expected 3% annual inflation rate for the US dollar, the loans are expected to carry a nominal interest rate of at least 12%. The real rate of return on equity for this type of investment is taken as 12%. Therefore, the weighted average real cost of capital financed through 70% borrowed money and 30% equity financing is calculated at 10%. The domestic annual inflation rate in North Cyprus is assumed to be 80%, and the end-of-1998 exchange rate is set at 290 050 TRL = 1 USD (in 1999, 429 900 Turkish lira [TRL] = 1 United States dollar [USD]) (Table 5).

Analysis results

Various unit costs of water

The objective of this part of our feasibility study is to estimate the minimum that must be charged per cubic metre of water to make water shipment by tanker from Turkey to North Cyprus feasible. This is a function of (1) the costs of the project; and (2) the efficiency of authorities in managing the water systems. The real net cash flow constructed from the owner's point of view enables us to derive the financial cost per cubic metre of water. The cost per cubic metre of water, computed at various stages of the delivery process, is the break-even average real price evaluated at the implementation stage of the project (December 1998) from the equity point of view, using a 12% real discount rate. The first calculated cost per cubic metre of water is the cost of transportation, which excludes installation costs at both ends, leakage in the system, and ongoing financial management (delays in reading the meter, billing, and payments, and adjusting water tariffs for inflation). The cost of transporting the water is found to be 0.46 USD/m^3; this figure does not include payment for raw water to Turkey. As a comparison, however, the cost

Table 5. Exchange rates, inflation rates, and financing.

Inflation and exchange rates			
Domestic inflation rate (%)			80
US inflation rate (average over last 5 years) (%)			3.0
Real exchange rate (TRL/USD) (149 000 × 1.8/1.03)			290 050 (year end 1998)
Financing (amounts)			
From Turkish or Turkish-guaranteed USD credit			
(70% of total investment costs) (USD)			11 707 500
From equity (30% of total investment costs) (USD)			5 017 500
	Real	Nom	
Interest rates (%)		inal	
Interest rate (%)	4.0		
Risk for Turkey (%)	5.0	—	
USD borrowing rate for Turkey (%)	9.0	—	
USD return on equity (%)	12	12.0	
		—	
Financing (terms)			
Percentage from equity (%)	30		
Percentage borrowed (%)	70		
Number of years for repayment	15		

Source: Data on the tanker project were obtained from the Undersecretary's and the Port General Directorate's offices of the Transportation Ministry, and data on the pipeline project were obtained from the Water Works Office of the Ministry of the Interior of the Turkish Republic of Northern Cyprus.
Note: TRL, Turkish lira (in 1999, 429 900 Turkish lira [TRL] = 1 United States dollar [USD]); USD, United States dollar.

of transporting 1 m^3 of water in water bags from Anamur to Kumkoy, a distance of 84 km, is estimated at 0.55 USD.[4]

The unit cost of water by tanker to Kumkoy increases to 0.79 USD/m^3 when the cost includes investment in the infrastructure required in North Cyprus (Table 6). This price also includes port handling charges in Turkey and operating costs in North Cyprus but excludes any payment to Turkey for the raw water (perhaps 0.15 USD/m^3), cost of leakage in the country's distribution system, and financial losses resulting from inefficient pricing or collection policies.

Leakage of water from the distribution system to households is also a cost. Adding the present 30% leakage, as well as unpaid deliveries, to the transportation

---

[4] See "Contract on Transporting Water from Turkey to the Turkish Republic of Northern Cyprus, Between the Mediterranean Water Supply A.S. (Mr. Akif Alpar) and the Ministry of the Interior of the TRNC, the Water Works Department (Mr. Mustafa Can)," 30 Dec 1997, p. 3, art. 3.

Table 6. Cost of water at various stages in the delivery process (USD/m^3).

Transportation cost of water	0.46
Cost of water to Kumkoy	0.79
Cost of water to households	
(20% leakage)	0.99
(30% leakage)	1.13

Source: Calculations from data in Tables 1–5.
Note: Costs do not include any payment for raw water to Turkey, which could be around 0.15 USD/m^3. USD, United States dollar.

and infrastructure costs, the cost of delivering 1 m^3 of water to households would be 1.13 USD/m^3, excluding any payment for water in Turkey and any water treatment costs (Table 6).

The above analysis involves an evaluation of the real net cash flow from the equity point of view. The cash-flow statements for selected cases expressed in real (1998) prices are shown in Tables 7 and 8.

Some financial management aspects in determining the unit cost of water

In North Cyprus, the Waterworks Department of the Ministry of the Interior has responsibility for distributing water to municipalities and other local authorities, repairing breakdowns, and general maintenance of the distribution system. Municipalities read the water meters, bill the customers, and collect payments to meet their own budgets, but they have not been very efficient at it. In an inflationary environment, lags in reading meters, billing, and payment have a great impact on the net cash flow of the utility. Meters are read every 2 months. In Gazimagusa, bills are filled in and given to customers on the spot, and consumers are expected to pay within the 2-month period before the meter is read again. A graduated surcharge is added for delays in payment. In Nicosia, rather than filling in the bill on the spot, it is prepared in the municipal office and brought to the consumer the next time that the meter is read, which results in a 2-month time lag in billing.

In this section, the cost of water is calculated for various scenarios, putting the following factors into relation: billing period, payment lag after billing, and frequency in adjusting nominal prices for inflation. For this analysis, rather than using annual cash flow, we construct monthly cash flow for any year of operation and use the value of sales at the end of 1998 to find the break-even price of water (P^*_0) that would yield equal revenues (in present-value terms) under the various

Table 7. Cash flow statement (1998 prices) total investment point of view (million TRL).

	1998	1999	2000	2005	2010	2013	2014
Inflation index	1.00	1.80	3.24	61.22	1 156.83	6 746.64	12 143.95
Receipts							
Sales revenue	0	1 927 368	1 927 368	1 927 368	1 927 368	1 927 368	0
Change in accounts receivable	0	0	0	0	0	0	0
Liquidation value							
Tanker	0	0	0	0	0	0	278 283
Mooring system	0	0	0	0	0	0	69 571
Boat	0	0	0	0	0	0	8 696
Sea pipeline	0	0	0	0	0	0	58
Land pipeline	0	0	0	0	0	0	58
Reservoir (Kumkoy)	0	0	0	0	0	0	208
New pumps (Serhatkoy)	0	0	0	0	0	0	33
Cash inflow	0	1 927 368	1 927 368	1 927 368	1 927 368	1 927 368	356 907
Expenditures							
Investment costs							
Tanker	2 782 835	0	0	0	0	0	0
Mooring system	695 709	0	0	0	0	0	0
Boat	86 964	0	0	0	0	0	0
Sea pipeline	391 336	0	0	0	0	0	0
Land pipeline	347 854	0	0	0	0	0	0
Reservoir (Kumkoy)	1 252 276	0	0	0	0	0	0
New pumps (Serhatkoy)	347 854	0	0	0	0	0	0
Total investment costs	5 904 828	0	0	0	0	0	0

(continued)

Table 7. Concluded.

	1998	1999	2000	2005	2010	2013	2014
Operating costs							
Crew salaries	0	176 327	181 225	207 833	238 348	258 767	0
Boat staff salaries	0	27 457	28 220	32 363	37 115	40 295	0
Insurance	0	55 657	55 657	55 657	55 657	55 657	0
Maintenance	0	51 222	51 222	51 222	51 222	51 222	0
Fuel and oil	0	453 924	453 924	453 924	453 924	453 924	0
Manavgat handling charges	0	304 647	304 647	304 647	304 647	304 647	0
Miscellaneous	0	8 922	8 922	8 922	8 922	8 922	0
Annual cost of water	0	0	0	0	0	0	0
Total operating costs	0	1 078 156	1 083 817	1 083 817	1 114 568	1 173 433	0
Working capital							
Change in accounts payable		-60 686	-26 971	-26 971	-26 971	-26 971	33 714
Change in cash balances		44 308	19 925	20 506	21 172	21 617	-26 791
Total change		-16 738	-7 046	-6 466	-5 800	-5 354	6 924
Cash outflow	5 904 828	1 061 778	1 076 770	1 108 102	1 144 034	1 168 079	6 924
Net cash flow	-5 904 828	865 590	865 598	819 266	783 334	759 289	349 983

Source: Calculations from data in Tables 1–5.
Note: TRL, Turkish lira (in 1999, 429 900 Turkish lira [TRL] = 1 United States dollar [USD]); USD, United States dollar.

Table 8. Cash flow statement (1998 prices) owner's point of view (million TRL).

Year	1998	1999	2000	2005	2010	2013	2014
Loan inflow	4 072 505						
Net cash flow before financing	−5 904 828	865 590	850 598	819 266	783 334	759 289	349 983
Net debt-financing cash flow	0	−650 190	−631 252	−544 524	−469 711	−429 852	0
Net cash flow after financing	−1 832 323	215 400	219 345	274 742	313 623	329 437	349 983
NPV	−0						
Equity return rate (real)	12%						

Source: Calculations from data in Tables 1–5.
Note: NPV, net present value; TRL, Turkish lira (in 1999, 429 900 Turkish lira [TRL] = 1 United States dollar [USD]); USD, United States dollar.

scenarios. The first year's revenues in the annual formulation of the model are given in equation [1]:

$$\frac{P_o \times (1 + gP_A) \times Q}{(1 + gP_A) \times (1 + r_A)} \qquad [1]$$

where P_o is the end-of-year price of water for 1998 (break-even price for the initial year) obtained from the annual net cash flow (estimated at 0.7915 USD, when the NPV is set to zero); Q is the amount of water sold in a year; gP_A is the annual inflation rate (80%); and r_A is the annual discount rate, which is the real rate of return on equity (12%). For the case under study, the value of equation [1] is 4 970 462 USD in the first year of operation.

In the first set of scenarios, it is assumed that there is monthly billing and instantaneous adjustment in the prices for inflation, and the payment lag after billing is allowed to vary as "no lag," "1-month lag," and "3-month lag" in payment after billing. For equivalence between the value of annual cash flows and monthly cash flows yielding annual revenues for any given year, we have equation [2], calculating the break-even prices of water when payments are made more frequently than once a year:

$$\frac{\sum\limits_{i=1}^{B_c} \{[P^*_o \times (1 + gP_m)^{NPA_\Pi - P_L}] \times Q/B_c\}}{\sum\limits^{B_c} \{[(1 + gP_m)^{(12/B_C) \times k}] \times (1 + r_m)^{(12/B_C) \times Q/B_C}\}} = \frac{P_o \times (1 + gP_A) \times Q}{(1 + gP_A) \times (1 + r_A)} \qquad [2]$$

where P^*_o is the initial price if payments are made monthly; gP_m is the monthly inflation rate; r_m is the monthly discount rate; NPA_Π is the period representing the month that the adjustment of nominal price for inflation is made; P_L is the payment lag after billing expressed in number of months; B_c is the number of billing cycles in a year; n is $(12/B_c) \times k$; and k refers to the particular billing cycle in the year (see Table 9).

Table 9. Value of water for alternative frequencies in adjusting nominal prices for inflation.

						Month						
	1	2	3	4	5	6	7	8	9	10	11	12
For instantaneous adjustment												
W	1	2	3	4	5	6	7	8	9	10	11	12
For quarterly adjustment												
W	1	1	1	4	4	4	7	7	7	10	10	10
For semiannual adjustment												
W	1	1	1	1	1	1	7	7	7	7	7	7
For annual adjustment												
W	1	1	1	1	1	1	1	1	1	1	1	1

Source: Calculations from data in Tables 1–5.
Note: Based on equation 2, where P_L is equal to 1 (1-month payment lag after billing), 2 (2-month payment lag after billing), and 3 (3-months payment lag after billing); and B_c is equal to 12 (monthly billing), 4 (quarterly billing), and 2 (semiannual billing).

We now solve equation [2] for the value of P^*_o, which is the initial real price that must be set at the end of 1998 for the periodic system of payments to yield the same revenue in present-value terms as obtained under the assumption that water is all used and all sold at the end of each year. This analysis is applicable to all sources of water; it is not just a feature of the tanker project. It is equally applicable to water obtained from wells, dams, water bags, pipelines, or desalination plants. The results obtained are given in Table 10.

At a zero rate of leakage in the distribution system, with billing carried out monthly, no lag in payment, and instantaneous adjustment of price for inflation, the break-even price is 0.751 USD. In the event of 1 or 3 months of payment lag after billing, the break-even price of water rises to 0.789 USD and 0.870 USD, respectively, because of the time value of money. Households are equally well off financially if they pay 0.751 USD/m^3 with no payment lag, 0.789 USD/m^3 with a 1-month payment lag, or 0.870 USD/m^3 with a 3-month payment lag.

Results from the annual cash-flow statements are used to determine the equivalent break-even price for billing periods of 1 and 2 months when nominal prices are adjusted instantaneously for inflation. The payment lag after billing is taken also with "zero," "1-month," and "3-month" lags after billing. The results

Table 10. Break-even real prices of water per cubic metre for various scenarios from end of 1998.

Billing cycle	Payment lag after billing	Time lag to adjust for inflation	Break-even price of water at various levels of leakage (USD)			
			0%	10%	20%	30%
Annual	0	Instantaneous	0.7915	0.8795	0.9894	1.1308
Monthly	0	Instantaneous	0.7510	0.8345	0.9388	1.0730
Monthly	1 month	Instantaneous	0.7887	0.8764	0.9860	1.1269
Monthly	2 months	Instantaneous	0.8283	0.9204	1.0354	1.1834
Monthly	3 months	Instantaneous	0.8699	0.9666	1.0874	1.2428
Monthly	1 month	Quarterly	0.8274	0.9194	1.0343	1.1821
Monthly	1 month	6 months	0.8872	0.9858	1.1090	1.2675
Monthly	1 month	Annually	1.0124	1.1250	1.2656	1.4464
Monthly	2 months	Quarterly	0.8690	0.9656	1.0863	1.2415
Monthly	2 months	6 months	0.9317	1.0353	1.1647	1.3311
Monthly	2 months	Annually	1.0633	1.1815	1.3291	1.5191
Monthly	3 months	Quarterly	0.9126	1.0140	1.1408	1.3038
Monthly	3 months	6 months	0.9785	1.0873	1.2231	1.3980
Monthly	3 months	Annually	1.1166	1.2408	1.3958	1.5953
2 months	1 month	Instantaneous	0.7925	0.8806	0.9906	1.1322
2 months	2 months	Instantaneous	0.8323	0.9248	1.0404	1.1890
2 months	3 months	Instantaneous	0.8740	0.9712	1.0926	1.2487
2 months	1 month	Quarterly	0.8314	0.9238	1.0392	1.1877
2 months	1 month	6 months	0.9139	1.0155	1.1424	1.3056
2 months	1 month	Annually	1.0429	1.1588	1.3037	1.4900
2 months	2 months	Quarterly	0.8731	0.9702	1.0914	1.2474
2 months	2 months	6 months	0.9598	1.0665	1.1997	1.3712
2 months	2 months	Annually	1.0952	1.2170	1.3691	1.5648
2 months	3 months	Quarterly	0.9169	1.0189	1.1462	1.3100
2 months	3 months	6 months	1.0079	1.1200	1.2599	1.4400
2 months	3 months	Annually	1.1502	1.2781	1.4378	1.6433

Source: Calculations from data in Tables 1–5.
Note: USD, United States dollar.

are presented in Table 10. It was found that billing for water consumption every month, rather than once every 2 months, does not have a great impact on the price of water, less than 1% per cubic metre, or 0.789 USD versus 0.793 USD/m^3 (assuming a zero level of leakage from the distribution system).

Billing every 2 months and getting paid with a 2-month lag is now the case in Gazimagusa, except that at present, nominal prices are not adjusted to inflation instantaneously but annually. Billing every 2 months but getting paid after 3 months describes the application for Nicosia, where the break-even price is 1.150 USD/m^3, with annual adjustment for inflation, which is 0.055 USD higher than the break-even price for Gazimagusa (1.095 USD).

The break-even prices given here for Nicosia and Gazimagusa are based on the assumption of no leakage from the distribution systems. However, in informal communication with the authors, local officials involved in dealing with the issue in North Cyprus estimated leakage at 25–30%. If we factor in 30% leakage, households in Gazimagusa would have to pay 1.565 USD, as opposed to 1.095 USD; and in Nicosia, 1.643 USD, rather than 1.150 USD/m^3 of water, substantially higher than the real landed cost of 0.751 USD, delivered at Kumkoy by tanker.

Sensitivity analysis

To determine the effects of an investment cost overrun and the rate of return on equity on the outcome of the project, a sensitivity analysis was carried out. Investment costs may go up because of a rise in the cost of inputs; the amount of physical inputs may increase; or there may be delays in completing construction. Table 11 gives the break-even prices of water per cubic metre for various levels of the NPVs of investment cost overruns. It was found that the break-even annual price of water is somewhat sensitive to investment cost overruns: a 20% increase in the investment cost results in the real price of water rising from 0.792 USD/m^3 to 0.868 USD/m^3, about a 10% increase in price. However, price is not nearly as sensitive to cost overruns as it is to water leakage in the system. Table 10 shows that a 10% level of water leakage would require water prices to rise by 10%, and a 20% level of leakage would cause the price to rise by 20%.

The required rate of return on equity is another factor that may affect the outcome of the project. The sensitivity analysis of this variable on the break-even price of water is given in Table 12, which shows that the project is sensitive to

Table 11. Sensitivity analysis of investment cost overruns on the break-even price of water.

%	USD
−0.20	0.7147
−0.15	0.7339
−0.10	0.7531
−0.05	0.7723
0.00	0.7915
0.05	0.8107
0.10	0.8300
0.15	0.8492
0.20	0.8684

Source: Calculations from data in Tables 1–5.
Note: USD, United States dollar.

Table 12. Sensitivity analysis of real rate of return on equity on the break-even price of water.

%	USD
10	0.7782
12	0.7915
14	0.8052
16	0.8191
18	0.8332
20	0.8476

Source: Calculations from data in Tables 1–5.
Note: USD, United States dollar.

the required rate of return as well, but less so than to investment cost overruns or leakage. If the required rate of return is raised from a real rate of 12% to a real rate of 20% (66% increase), the required increase in the price of water is about 6%.

Conclusion

To solve the water-shortage problem in North Cyprus, various projects are planned for potential implementation. Conversion of traditional irrigation methods to modern irrigation on 10 000 dn in Guzelyurt and the rehabilitation and use of treated wastewater from Haspolat Wastewater Treatment Plant for agriculture are two projects expected to be completed by the end of 1998. In this study, a financial feasibility analysis of importing 7 Mm^3 of water to North Cyprus from Turkey by a tanker was carried out. Even more importantly, an analysis of alternative pricing policies was formulated, reflecting the various management practices of water-resource authorities in North Cyprus.

The transportation cost per cubic metre of water imported from Manavgat to Kumkoy by a tanker with a capacity of 40 000 m^3 was found to be on average $0.46 USD/$m^3$. This price does not include any infrastructure to be built in North Cyprus, port handling charges in Turkey, or payment for water to Turkey. When infrastructure and operating costs in North Cyprus and port handling charges are included, the cost of water delivered to Kumkoy is expected to be 0.79 USD/m^3. This price also excludes any payment to Turkey for the raw water. These results indicate that water-tanker transportation between Turkey and North Cyprus is highly competitive with other methods of supply, such as desalination, which cost at least 50% more (Rogers 1994).

A monthly net cash-flow statement was used to analyze the effects of various financial aspects on the price of water to the consumer, a method that is applicable to all sources of water supply. In this analysis, break-even prices were calculated to reflect the time value of money (households would be indifferent to this) in present-value terms. It has been observed that billing monthly or billing every 2 months does not significantly affect the price of water to the consumer. However, billing every 2 months with a payment lag of 2 months after billing (the case of Gazimagusa), or a payment lag of 3 months after billing (the case of Nicosia), combined with annually adjusted nominal water prices for inflation, affects the break-even price of water substantially, causing it to rise to 1.095 USD and 1.150 USD, respectively.

By far the most important variable determining the real price of water is the amount of leakage in the system. This variable is directly related to the management and maintenance practices of local water authorities. When water leakage

of 30% is taken into consideration, the break-even price of water increases to
1.565 USD for Gazimagusa and 1.643 USD for Nicosia. The model also enables
us to predict the price of water if the percentage of leakage is reduced from 30%
to 20% and 10%. Under the same circumstances, at a 20% leakage level, the price
of water in Gazimagusa and Nicosia would fall to 1.369 USD and 1.438 USD,
respectively.

A sensitivity analysis carried out on the impact of investment cost overruns
and the required rate of return on equity on the break-even price of water showed
that they affect the outcome of the project and, therefore, the cost of water per
cubic metre as well, but they are not as significant as poor water-management
practices that account for high rates of leakage in the distribution system and less
than efficient billing systems.[5]

References

Biçak, H. 1996. investment appraisal project of importing water from Turkey to Cyprus
by a tanker and through pipes. Harvard University, Investment Appraisal and Management
Program, Harvard Institute for International Development, Cambridge, MA, USA.

Biçak, H.; Özdemirag A. 1997. Estimation of forecasting of water demand and supply in
North Cyprus: the impact of various projects under various scenarios. Paper presented at
the Economic Research Seminar Series, Department of Economics, Eastern Mediterranean
University, Gazimagusa, North Cyprus.

Biyikoglu, G. 1995. Rainfall analysis in the Turkish Republic of Northern Cyprus.
Proceedings of the Second Water Congress, 23–24 Feb 1995. Cyprus Turkish Engineer
and Architecture Chamber Association, Nicosia, North Cyprus. [In Turkish]

Harberger, A.C.; Jenkins, G.P. 1998. Manual: cost–benefit analysis of investment
decisions. Program on Investment Appraisal and Management, Harvard Institute for
International Development, Cambridge, MA, USA.

Kibris. 1998a. 1997. President Suleyman Demirel brought the pearce water and returned.
Kibris, Nicosia, 26 Jul. [In Turkish]

---

[5] This paper was delivered in October 1998, and, therefore, December 1998 is referred
to in the future. At the time of writing, however, April 1999, the project still has not been
implemented. But it remains a serious alternative to transporting water in Medusa bags, because
existing sea installations, with a little alteration, will permit water transportation by tanker.

———— 1998b. Haspolat water will flow for agriculture. Interview made with the Mayor of Nicosia and the expert in charge of the project. Kibris, Nicosia, 15 Jul. Interview with the Turkish Mayor of Nicosia and the expert in charge of the project. [In Turkish]

MOAF (Ministry of Agriculture and Forestry). 1997a. Agricultural structure and production 1975–1995, Nicosia, North Cyprus.

———— 1997b. Agricultural structure and production 1996, Nicosia, North Cyprus.

Numan, T.; Agiralioglu, N. 1995. Forecasts and suggestions for the water problem of North Cyprus in the short run and long run. *In* Altay, G.; Borekci, O.; Akkoyunlu, A.; Altinbilek, D.; Gucbilmez, D., ed., Developments in the Civil Engineering, Proceedings of the Second Technical Congress, 18–20 Sep 1995, Istanbul, Turkey.

Özdemirag, A. 1998. Investment appraisal of a dam to be built in Yesilirmak. Department of Economics, Eastern Mediterranean University, Gazimagusa, North Cyprus. MSc Thesis.

Rogers, P. 1994. The agenda for the next thirty years. *In* Rogers, P. ed., Water in the Arab world: perspectives and prognoses. Harvard University Press, Cambridge, MA, USA.

Sevki M. 1997. Big Agricultural Project in Guzelyurt. Kibris, Nicosia, North Cyprus, 20 Nov. [In Turkish]

Chapter 8

Trends in Transboundary Water Resources: Lessons for Cooperative Projects in the Middle East

Aaron T. Wolf

Introduction

The 261 international rivers, covering almost one-half of the total land surface of the globe, provide ample opportunity for political tensions. Such has been the case in Africa, the Middle East, and Southeast Asia. Given water's preeminence as a critical resource and the fact that management of water resources is very poorly defined in the international arena, it is of little surprise that water and war are two topics assessed together with increasing frequency.

The history of hydropolitics along the rivers of the Middle East exemplifies both the worst and the best of international relations over water. All of the countries and territories riparian to the Jordan River — Israel, Jordan, the Palestine Authority, and Syria — are currently using between 95% and more than 100% of their annual renewable freshwater supply. In recent dry years, water consumption has routinely exceeded annual supply, and the difference has usually been made up through overdrafts on fragile groundwater systems. By 2020, water shortages will be the norm. Projected water requirements for 2020 are 2 000 Mm^3/year, or about 130% of current renewable supplies, for Israelis; 1 000 Mm^3, or 120% of current supplies, for Jordanians; and 310 Mm^3, or 150% of current supplies, for Palestinians on the West Bank and Gaza. The resolution of this crisis is extremely difficult because intense and fluctuating geopolitical forces have crafted political boundaries in direct contradiction to the natural boundaries of the watersheds in the region.

Although shared water resources have led to, and occasionally crossed, the brink of armed conflict, they have also been a catalyst for cooperation between otherwise hostile neighbours, albeit rarely and secretively. For example, despite a growing literature suggesting that Arab–Israeli warfare has had a "hydrostrategic" component, the evidence suggests that water resources were not at all factors in strategic planning during the hostilities of 1948, 1967, 1978, or 1982. The decision to go to war and strategic decisions made during the fighting, including the question of which territory it was necessary to capture, were not influenced by water scarcity or the location of water resources. Moreover, although questions of water allocation and rights have been among the most difficult components in the Arab–Israeli peace talks and a large number of studies have identified hydrostrategic territory and advised its retention, no territory to date has been retained simply because of the location of water. Solutions, in each case, have focused on creative joint management of the resource, rather than sovereignty.

Water and conflict[1]

A growing literature describes water both as an historic and, by extrapolation, a future cause of interstate warfare. Westing (1986) suggested that "competition for limited ... fresh water ... leads to severe political tensions and even to war." Gleick (1993) described water resources as military and political goals, using the Jordan and Nile as examples. Remans (1995) used case studies from the Middle East, South America, and South Asia as "well-known examples" of water as a cause of armed conflict. Samson and Charrier (1997) wrote that "a number of conflicts linked to fresh water are already apparent" and suggested that "growing conflict looms ahead." Butts (1997) suggested that "history is replete with examples of violent conflict over water" and named four Middle Eastern water sources particularly at risk. Finally, Homer-Dixon (1994), citing the Jordan and other water disputes, came to the conclusion that "the renewable resource most likely to stimulate interstate resource war is river water."

---

[1] The next two sections are drawn from Wolf, A.T. (1999b).

A close examination of the case studies cited in this literature reveals looseness in classification. Samson and Charrier (1997), for example, listed 18 cases of water disputes, only one of which was described as "armed conflict," and that particular case (on the Cenepa River) turned out not to be about water at all but about the location of a shared boundary, which happened to coincide with the watershed. No armed conflicts occurred in any of Remans' (1995) "well-known" cases (except the one between Israel and Syria, described below), nor in any of the other lists of water-related tensions presented.

The examples most widely cited are wars between Israel and its neighbours. Westing (1986) listed the Jordan River as a cause of the 1967 war and, in the same volume, Falkenmark (1986), mostly citing Cooley (1984), described water as a causal factor in both the 1967 war and the 1982 Israeli invasion of Lebanon. Myers (1993), using water in the Middle East as his first defining example of "ultimate security," wrote that "Israel started the 1967 war in part because the Arabs were planning to divert the waters of the Jordan River system." In fact, after Israel's 1982 invasion of Lebanon, a "hydraulic-imperative" theory was developed in academic literature and the popular press (see, for example, Davis et al. 1980; Stauffer 1982; Schmida 1983; Stork 1983; Cooley 1984; Dillman 1989; and Beaumont 1991). This theory describes the quest for water resources as the motivator for Israeli military conquests, both in Lebanon in 1979 and 1982 and earlier, on the Golan Heights and West Bank in 1967.

The main problem with these theories is a complete lack of evidence. Although shots were fired over water conflict between Israel and Syria in 1951–53 and 1964–66, the final exchange, including both tanks and aircraft on 14 July 1966, stopped Syrian construction of the diversion project in dispute, effectively ending water-related tensions between the two states. The 1967 war broke out almost a year later, and no link has ever been documented. The 1982 invasion provides even less evidence of any relation between hydrologic and military decision-making. In extensive papers investigating precisely such a linkage between hydro- and geostrategic considerations, both Libiszewski (1995) and Wolf (1995a) concluded that water was neither a cause nor a goal of any Arab–Israeli warfare.

To be fair, it should be noted that this analysis only describes the relationship between interstate armed conflict and water resources as a scarce resource. Internal disputes, such as those between interests or provinces, as well as those in which water was a means, method, or victim of warfare, are excluded. Also excluded are disputes in which water is incidental to the dispute, such as those about fishing rights, access to ports, transportation, or river boundaries. Many of the authors — notably, Gleick (1993), Libiszewski (1995), and Remans (1995) — are very careful about these distinctions. The bulk of the articles cited above, then, turn out to be about political tensions or stability, rather than about warfare, or about water as a tool, target, or victim of armed conflict — all important issues, just not the same as "water wars."

To cut through the prevailing anecdotal approach to the history of water conflicts, the analysts working on the database project investigated those cases of international conflict in which armed exchange was threatened or took place over water resources per se. We used the most systematically collected information available on international conflict — the International Crisis Behavior data set, collected by Jonathan Wilkenfeld and Michael Brecher (1997). This data set contains information only on disputes the head researchers considered international crises. Their definition of an international crisis was any dispute in which (1) basic national values are threatened (for example, territory, influence, or existence); (2) time for making decisions is limited; and (3) the probability of military hostilities is high. Using these guidelines, they identified 412 crises for the period of 1918–94. Of these, only 7 were even partially related to water resources (Figure 1). Thus, the actual history of armed conflict over water is somewhat less dramatic than the water wars literature would lead one to believe: a total of 7 incidents. In 3 of these, no shots were fired. As near as we can find, there has never been a single war fought over water.[2]

This is not to say that there is no history of water-related violence — quite the opposite is true — only that these incidents are at the subnational level, generally between tribes, water-use sectors, or provinces. Nationally internal water conflicts are, in fact, quite prevalent. Interprovincial violence along the Cauvery River in India is but one example. California farmers blew up a water pipeline meant for Los Angeles. Much of the violent history in the Americas between indigenous peoples and European settlers has included struggles over water. The US desert

---

[2] This in not quite true. The earliest documented interstate conflict known is a dispute between the Sumerian city-states of Lagash, and Umma over the right to exploit boundary channels along the Tigris in 2500 BCE (Cooper 1983). In other words, the last and only water war was 4 500 years ago.

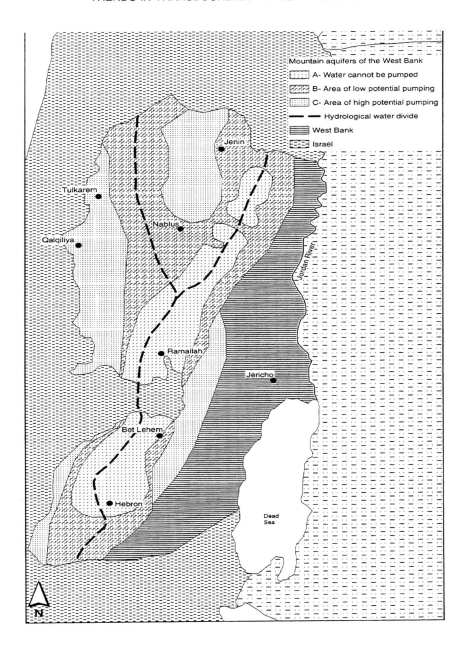

Figure 1. Mountain aquifers of the West Bank.

Cases of acute water-related disputes[3]

1948 — Partition between India and Pakistan leaves the Indus Basin divided in a particularly convoluted fashion. Disputes over irrigation water exacerbate tensions in the still-sensitive Kashmir region, bringing the two riparians "to the brink of war." Twelve years of World Bank-led negotiations lead to the 1960 Indus Waters Agreement.

February 1951 – September 1953 — Israel and Syria exchange sporadic fire over Israeli water development works in the Huleh Basin, which lies in the demilitarized zone between the two countries. Israel moves its water intake to the Sea of Galilee.

January–April 1958 — Amid pending negotiations over the Nile waters, Sudanese general elections, and an Egyptian vote on Sudan–Egypt unification, Egypt sends an unsuccessful military expedition into territory in dispute between the two countries. Tensions were eased (and a Nile Waters Treaty signed) when a pro-Egyptian government was elected in Sudan.

June 1963 – March 1964 — The 1948 boundaries left Somali nomads under Ethiopian rule. Border skirmishes between Somalia and Ethiopia took place over disputed territory in Ogaden desert, which includes some critical water resources (both sides were also aware of oil resources in the region). Several hundred were killed before cease-fire was negotiated.

March 1965 – July 1966 — Israel and Syria exchanged fire over "all-Arab" plan to divert the Jordan River headwaters, presumably to preempt the Israeli National Water Carrier, an out-of-basin diversion plan from the Sea of Galilee. Construction of the Syrian diversion was halted in July 1966.

April–August 1975 — In a particularly low-flow year along the Euphrates (owing to filling of upstream dams), Iraq claimed that the flow reaching its territory was "intolerable" and asked that the Arab League intervene in its dispute over this with Syria. The Syrians claimed that less than half the river's normal flow was reaching its borders that year, and after a barrage of mutually hostile statements, an Arab League technical committee that had been formed to mediate the conflict pulled out. In May 1975, Syria closed its airspace to Iraqi flights, and both Syria and Iraq reportedly transferred troops to their mutual border. Only mediation on the part of Saudi Arabia broke the increasing tension.

April 1989 – July 1991. Two Senegalese peasants were killed over grazing rights along the Senegal River, which forms the boundary between Mauritania and Senegal. This sparked smoldering ethnic and land-reform tensions in the region. Several hundred people were killed as civilians from border towns on either side of the river attacked each other, until each country used its army to restore order. Violence broke out sporadically until diplomatic relations were restored later in 1991.

---

[3] I define an acute dispute as one involving the mobilization of armies or shots fired in an international setting.

state of Arizona even commissioned a navy (made up of one ferry boat) and sent its state militia to stop a dam and diversion project on the Colorado River in 1934 (Fredkin 1981).

One need look no further than relations between Bangladesh and India to note that internal instability can both be caused by and exacerbate international water disputes. At one time, India built a barrage at Farakka to divert a portion of the Ganges away from its course into Bangladesh toward Calcutta, 100 miles (1 mile = 1.609 km) to the south, for the purpose of flushing silt away from Calcutta's seaport. Adverse effects in Bangladesh, resulting from reduced upstream flow, included degradation of both surface and groundwater, change in morphology, impeded navigation, increased salinity, degraded fisheries, and danger to water supplies and public health. Environmental refugees from the affected areas further compounded the problem. Ironically, many of those displaced in Bangladesh found refuge in India (Biswas and Hashimoto 1996).

So, although no water wars have occurred, there is ample evidence that the lack of fresh water has lead to occasionally intense political instability and that on a small scale, acute violence can result. What we seem to be finding, in fact, is that geographic scale and intensity of conflict are inversely related.

Water and cooperation

The history of water dispute resolution, in contrast to that of conflict, is much more impressive. The Food and Agriculture Organization of the United Nations has identified more than 3 600 treaties relating to international water resources, dating between 805 and 1984, the majority of which deal with some aspect of navigation (FAO 1978, 1984). Since 1814 about 300 international treaties have been negotiated to deal with nonnavigational issues of water management: flood control, hydropower projects, and allocations for consumptive or nonconsumptive uses in international basins. Restricting ourselves to those signed in this century that deal with water per se and excluding those that deal with boundaries or fishing rights, we have collected the full texts of 149 treaties and have scanned or digitally entered them into our Transboundary Freshwater Dispute Database at Oregon State University, in conjunction with projects funded by the World Bank and the US Institute of Peace. Negotiating notes and published descriptions of many treaty negotiations are also being collected.

Some 14 case studies have been described in detail and in similar format for purposes of comparison in forthcoming research work. These cases include nine watersheds: the Danube, Euphrates, Jordan, Ganges, Indus, La Plata, Mekong, Nile, and Salween; two aquifer systems: US–Mexico shared systems and the West

Bank aquifers; two lake systems: the Aral Sea and the Great Lakes; and one engineering endeavour: the Lesotho Highlands Project. Jesse Hamner, now at Emory
University, developed a systematic database compilation of these treaties, creating
fields for the inclusion of basins, countries involved, dates signed, treaty topics,
allocation measures, conflict resolution mechanisms, and nonwater linkages. Analyses from this database are described in greater detail in Wolf (1997) and in Hamner and Wolf (1998). Details of the fourteen case studies listed can be found in
Bingham et al. (1994).

The historic reality has been quite different from what the water wars literature would have one believe. In modern history, only seven minor skirmishes
have occurred over international waters — invariably, other interrelated issues also
factor in. Conversely, more than 3 600 treaties have been signed, historically, over
different aspects of international waters, almost 150 in this century dealing with
water qua water, many showing tremendous elegance and creativity for dealing
with this critical resource. This is not to say that armed conflict has not taken
place over water, only that such disputes generally are between tribes, water-use
sectors, or provinces. A close look at the very cases most commonly cited as
water conflicts reveals ongoing dialogue, creative exchanges, and negotiations
leading fairly regularly to new treaties. A new question emerges that is arguably
more thought provoking and less dramatic than that of where the next water war
will break out: Given all of the seemingly conflict-inducing characteristics of
transboundary waterways, why has so little international violence taken place?

Water and the Arab–Israeli peace negotiations[4]

The history of hydropolitics along the rivers of the Middle East has roots older
than the states themselves. Water-related conflict, for example, informed the
borders of the British and French mandates, later the modern entities of Israel, Jordan, Lebanon, the Palestine Authority, and Syria. As each of these entities developed its water resources unilaterally, dispute became inevitable — every state or
territory in the Jordan River watershed has at least some of its water sources in
a different and occasionally hostile state or territory. Exchanges of fire actually
broke out between Israel and Syria over water in the mid-1950s and 1960s. The
problems were only exacerbated with the 1967 war.

The West Bank overlies three major aquifers, two of which Israel has been
tapping into from its side of the Green Line since 1955. In the years of Israeli
occupation, a growing West Bank and Gaza population, along with burgeoning

---

[4] This section draws from Wolf (1999a).

Jewish settlements, increased the burden on the limited groundwater supply, resulting in an exacerbation of already tense political relations. Palestinians have objected strenuously to Israeli control of water resources and development of settlements, which they see as being at their territorial and hydrologic expense, whereas Israeli authorities view hydrologic control in the West Bank as defensive. With about 30% of Israeli water originating on the West Bank, the Israelis perceive the need to limit groundwater exploitation in these territories to protect the resources themselves and their wells from saltwater intrusion.

Because of the disparate depths needed to reach water from these aquifers in the coastal plain and in the Judean hills (about 60 m in the plain, 150–200 m in the foothills, and 700–800 m in the hills [Goldschmidt and Jacobs 1958; Weinberger 1991]) and the resulting cost differences in drilling and pumping wells in these areas, portions of the aquifers are especially vulnerable to overpumping along a narrow western band of a northern lobe of the West Bank, in the region of Kalkilya and Tulkarm (see Figure 1).

In 1977, the right-wing Likud Party gained control of the Israeli parliament for the first time. As Israeli Prime Minister Menachem Begin was preparing for negotiations with Egyptian President Anwar Sadat, he asked the Water Commissioner at the time, Menachem Cantor, to provide him with a map of Israeli use of water originating on the West Bank and guidelines as to where Israel might relinquish control if protecting Israel's water resources was the only consideration.

Cantor concluded that a "red line" could be drawn, beyond which Israel should not relinquish control, a line running north to south and following roughly the 100- to 200-m contour line along both "lobes" of the West Bank. Israeli water planners still refer to this red line as a frame of reference, and it has occasionally been included in academic boundary studies of the region (Area C in Figure 1). This concept of a red line was later expanded by others to areas of the northern headwaters and the Golan Heights.

In 1991, the Jaffee Center for Strategic Studies at Tel Aviv University asked two researchers — Yehoshua Schwartz, Director of Tahal, Israel's water planning agency, and Aharon Zohar, also at Tahal at the time — to undertake a study of the regional hydrostrategic situation and the potential for regional cooperation. The result, a 300-page document titled *Water in the Middle East: Solutions to Water Problems in the Context of Arrangements between Israel and the Arabs,* was one of the most comprehensive studies of its kind (Schwartz and Zohar 1991). It examined a number of scenarios for regional water development, including possible arrangements between Israel and Egypt, Iraq, Jordan, Lebanon, the Palestinians on the West Bank and Gaza, Saudi Arabia, Syria, and Turkey. Scenarios were included both for regional cooperation and for its absence. Evaluations were made of hydrologic, political, legal, and ideological constraints. The impacts of potential global climatic change were also considered. The study showed, in the words of Joseph Alpher, director of the Jaffee Center, "the potential beauty of multi-lateral negotiations" (Alpher 1994).

Some of the findings of the study contradicted government policies at the time. In the sections on possible arrangements between Israel and the Palestinians and between Israel and Syria, maps of the West Bank and Golan Heights included lines to which Israel might relinquish control of the water resources in each area without overly endangering its own water supply. The line in the West Bank, based on Cantor's red line, suggested that with legal and political guarantees, Israel might give Palestinian authorities control of the water resources in more than two-thirds of the West Bank. This would not threaten Israel's water sources from the Yarkon-Taninim (western mountain) aquifer, although the authors advocated relinquishing control beyond the red line.

The same was true of more than half of the Golan Heights. These maps contradicted the position of the Ministry of Agriculture. Headed by Rafael Eitan of the right-wing Tzomet party, the Ministry's position was that to protect Israel from threats to both the quantity and the quality of its water, it had to retain political control over the entire West Bank.[5]

On 12 December 1991, 70 copies of the report were sent throughout Israel for review, including to the Ministry of Agriculture. Calling the maps mentioned "an outline for retreat," Rafael Eitan and Dan Zaslavsky, whom Eitan had recently appointed Water Commissioner, insisted on a recall of the review copies and a delay in the release of the report. In January 1992, the Israeli military censor backed the position of the Ministry of Agriculture and, citing the sensitivity of the report's findings, censored the report in its entirety.[6]

Bilateral and multilateral negotiations

The Gulf War in 1990 and the collapse of the Soviet Union caused a realignment of political alliances in the Middle East that finally made possible the first public, face-to-face peace talks between Arabs and Israelis at a meeting held in Madrid on 30 October 1991. During bilateral negotiations between Israel and each of its

---

[5] Eitan's position, argued in full-page ads in the Israeli press, has little bearing in hydrogeology, as discussed in Wolf (1995a).

[6] When peace talks began in 1991, the document remained censored, for fear its release would reveal Israeli negotiating strategy. The document has not been released to date.

neighbours, it was agreed that a second track be established for multilateral negotiations on five subjects deemed "regional," including water resources. These two mutually reinforcing tracks — the bilateral and multilateral — led to a peace treaty between Israel and Jordan and a declaration of principles for agreement between Israel and the Palestinian Authority. Both have a water component in terms of allocations and projects. In neither, however, has water had any influence on the discussions over final boundaries.

Israel–Jordan Peace Treaty

Israel and Jordan have probably had the warmest relations of any two states legally at war with each other. Communication between the two states has taken place since the creation of each, ameliorating conflict and facilitating conflict resolution on a variety of subjects, including water. The so-called Picnic Table Talks on allocations of the Yarmuk have taken place since the 1950s, and negotiations to formulate principles for water-sharing projects and allocations have occurred in conjunction with, and parallel to, both bilateral and multilateral peace negotiations.[7] These principles were formalized on 26 October 1994, when Israel and Jordan signed a peace treaty, ending more than four decades of a legal, when not actual, state of war.[8]

For the first time since the states came into being, the treaty legally defines mutually recognized water allocations. Acknowledging that "water issues along their entire boundary must be dealt with in their totality," the treaty spells out allocations for both the Yarmuk and Jordan rivers and Arava–Araba groundwater and calls for joint efforts to prevent water pollution. Recognizing "that their water resources are not sufficient to meet their needs," the treaty calls for ways of alleviating water shortage through cooperative projects, both regional and international.

The peace treaty also makes some minor boundary modifications. The Israel–Jordan boundary was delineated by Great Britain in 1922, following the centre of the Yarmuk and Jordan rivers, the Dead Sea, and Wadi Araba. In the late 1960s and 1970s, Israel had occasionally made minor modifications to the boundary south of the Dead Sea to make specific sections more secure from infiltrators. On occasion, this was also done to reach sites from which small wells might better be developed. In the last 16 years, no modifications have been made,

---

[7] For more details on the bilateral and multilateral talks on water, see Wolf (1995b).

[8] To my knowledge, these are the first international boundaries defined legally by Universal Transverse Mercator coordinates, as measured using the Global Positioning System.

except on the rare occasion when one of these local wells ran dry and had to be redug. All of these territorial modifications were reversed, and all affected land was returned to Jordan as a consequence of the peace treaty, although Israel still retains rights to use the water from these wells. Moreover, a small enclave of Jordanian territory in the Arava is being leased back to Israel in 25-year increments.

One other area was similarly affected. In 1926, a Jewish entrepreneur named Pinhas Rutenberg was granted a 70-year concession for hydropower generation at the confluence of the Yarmuk and Jordan rivers on land leased by Trans-Jordan. The dam that he built for this purpose would later be destroyed in the fighting of 1948, and the 1949 Armistice Line ended up leaving a small portion of Jordan under Israeli control. This land was farmed by the kibbutz Ashdot Ya'akov, established in 1933. With the 1994 peace treaty, sovereignty of the land was returned to Jordan, which in turn leased it back to Israel — Israeli kibbutznikim now travel into Jordanian territory regularly to farm their land.

In what will undoubtedly become a classic modification of the tenets of international law, Israelis and Jordanians, in their 1994 peace treaty, invented legal terminology to suit particularly local requirements. In negotiations leading up to the treaty, the Israelis, arguing that the entire region was running out of water, insisted on discussing only water "allocations," that is, the future needs of each riparian. Jordanians, in contrast, refused to discuss the future until past grievances had been addressed — they would not negotiate allocations until the historic question of water "rights" had been resolved.

There is little room to bargain between the past and the future, between rights and allocations. Negotiations had reached an impasse when one of the mediators suggested the term, "rightful allocations," to describe simultaneously historic claims and future goals for cooperative projects. This new term is now immortalized in the water-related clauses of the Israel–Jordan Peace Treaty.

Israeli–Palestinian Declaration of Principles and Interim Self-government Agreement

On 15 September 1993, Palestinians and Israelis signed the Declaration of Principles on Interim Self-Government Arrangement, which called for Palestinian autonomy in, and removal of Israeli military forces from, Gaza and Jericho. Among other issues, this bilateral agreement called for the creation of a Palestinian Water Administration Authority (later, the Palestinian Water Authority). Moreover, the first item in Annex III, on cooperation in economic and development programs, included a focus on

> Cooperation in the field of water, including a Water Development Program prepared by experts from both sides, which will also specify the mode of cooperation in the management of water resources in the West Bank and Gaza Strip, and will include proposals for studies and plans on water rights of each party, as well as on the equitable utilization of joint water resources for implementation in and beyond the interim period.

At about the same time, Israeli water managers discovered an additional 70 Mm^3/year of available yield in the eastern mountain aquifer — the only one of the three main West Bank units not being overpumped at the time. This probably did not hurt Jericho's choice as the first West Bank town to be given autonomy.[9]

Between 1993 and 1995, Israeli and Palestinian representatives continued negotiating toward a broadening of the interim agreement to encompass more West Bank territory. On 28 September 1995, the Israeli–Palestinian Interim Agreement on the West Bank and the Gaza Strip, known as Oslo II, was signed in Washington, DC. The issue of water rights was one of the most difficult to negotiate, and a final agreement was postponed, leaving water rights to be included in the negotiations for Final Status arrangements.[10] Nevertheless, a tremendous compromise was achieved by the two sides: Israel recognized the Palestinian claim to water rights, and a Joint Water Committee (JWC) was established to cooperate in management of West Bank water and to develop new supplies. The JWC, in principle, supervises joint patrols to investigate illegal water withdrawals; its first action was to discover and put a stop to illegal drilling in the Jenin area in December 1995 (*Israel Line*, 20 Dec 1995).[11]

In accordance with the agreement, Israeli forces withdrew from six Palestinian cities in order from north to south and from 450 towns and villages throughout the West Bank. The final status of Israeli settlements in the West Bank has yet to be determined. No territory at all was identified as being necessary for

---

[9] There is no evidence at all that the water was even considered in this choice; the comment is only this author's speculation.

[10] Oslo II estimates the future needs of West Bank Palestinians at 70–80 Mm^3/year. Until a final arrangement is negotiated, the two sides agree to cooperate to find a total 28.6 Mm^3/year for the interim period.

[11] E-mail summary of Israeli news, distributed by the Israeli Embassy in Washington, DC, USA. Unfortunately, the early promise of the JWC has not materialized. In the current political climate, in fact, it is all but inoperative.

Israeli annexation for access to water resources. The second and third cities scheduled for Israeli withdrawal — Tulkarm and Kalkilya — fell well within the red
line delineated in Israeli studies as needed to retain for water security.

This lack of correlation between transferred territory and the location of
water resources has become ever more apparent. Most recently, the November
1998 agreement reached at the Wye Plantation transferred an additional 13% of
the West Bank from Israeli to joint territory (3% of it as a nature preserve), and
14.2% that had been joint territory was moved to Palestinian control. Figure 2[12]
superimposes land transfers from both the Oslo II and the Wye negotiations onto
a map delineating the most hydrologically sensitive territories of the West Bank.
Even a cursory examination shows that hydrostrategic considerations are all but
ignored in favour of joint management and other creative solutions.

Negotiations between Israel, Lebanon, and Syria

At the time of writing, water has not been raised in official negotiations between
Israel and Syria.[13] Serious bilateral negotiations have only taken place since the
fall of 1995, and given the influence Damascus has on Beirut, Israel–Lebanon
talks are not likely until Israel and Syria make more progress. Israelis had hoped
to begin talks on water resources with the Syrians at a meeting in Maryland in
January 1996, but the Syrians reportedly refused to broaden the scope (*Israel Line*,
24 Jan 1996).

The basis for Israel–Syria negotiations is the premise of exchanging the
Golan Heights for peace. Discussions so far have focused on interpretations of
how much of the Golan, with what security arrangements, and for how much
peace. The crux of the territorial dispute is the question of which boundaries Israel
would withdraw — those between Israel and Syria have included the international
boundary between the British and French mandates (1923), the Armistice Line
(1949), and the cease-fire lines from 1967 and 1974 (Figure 3).

The Syrian position has been to insist on a return to the borders of 5 June
1967, whereas Israel refers to the boundaries of 1923. Although it has not been
mentioned explicitly, the difference between these two positions is precisely over
access to water resources. The only distinction between the two lines is the inclusion or exclusion of the three small areas constituting the demilitarized zone

---

[12] Many thanks to Robert Tobys, a geography student at Oregon State University, for
bringing his cartographic skills to bear on this intricate problem.

[13] In unofficial Track II discussions, water was the focus of meetings where Israelis
and Lebanese were present as early as 1993 and where Israelis and Syrians participated in
1994. Participants at these meetings did not necessarily have any official standing.

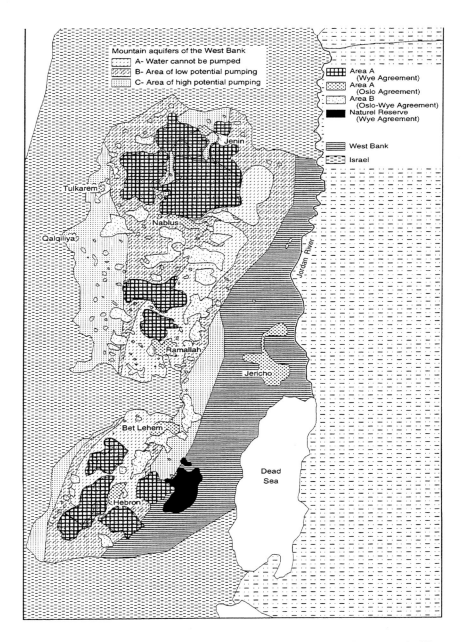

Figure 2. Mountain aquifers of the West Bank, with land transfers from Oslo II and the Wye agreements.

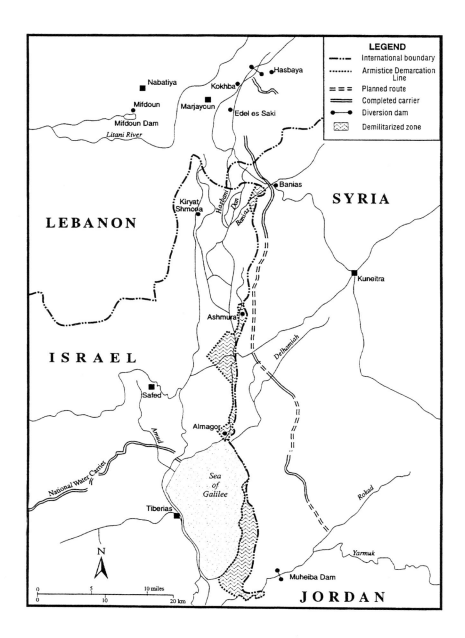

Figure 3. Boundaries between Israel, Jordan, Lebanon, and Syria.

between 1949 and 1967 — Givat Banias (the hill overlooking Banias springs), the Daughters of Jacob Bridge area, and the town of El-Hamma (Hamat Gader) — a total of about 60 km². These three territories were included in British Palestine specifically because of their access to the Jordan and Yarmuk rivers; moreover, as each is a relatively low-lying area with no strategic importance, their access to water is still considered their main value.[14]

Even before Israel–Syria negotiations began, a flurry of articles stressed the importance of water on the Golan Heights. As mentioned above, Schwartz and Zohar (1991) advised Israeli retention of the Golan Heights west of the Jordan River watershed line to guarantee continued control of the quantity and quality of water. In a 1994 study, Shalev (1994), himself a general retired from the Israeli army, cited five other retired generals on the importance of Israeli sovereignty over the Golan for the protection of water resources. Even in his small sample, Shalev finds a spectrum of opinion, from Major General Hofi, who suggested that Israel need retain a physical presence on the Golan Heights, to Major General Shafir, who advocated retaining at least the plateau above the Sea of Galilee, to former Chief-of-Staff Gur, who concluded that the water problem could be resolved politically in a peace treaty and that the territory was, therefore, not vital. Shalev concluded that Syria would not risk a war with Israel for water, especially as a diversion would take years to construct and would constitute a clear *causus belli*. It stands to reason, Shalev argued, that countries involved in water-sharing agreements would want to maintain them.

In the meantime, Schiff (1995), Tarnopolsky (1996), and others have argued in the popular Israeli and Jewish press that water's paramount importance may scuttle negotiations over the Golan, whereas Israeli politicians from the ruling Labour Party, including Prime Minister Shimon Peres and Foreign Minister Ehud Barak, argued that although the land may be negotiable, the water is not (*Jerusalem Post*, 6 and 27 Jan 1996).

Conclusion

Accounts of conflict related to water indicate that only seven minor skirmishes have occurred in this century and that no war has yet been fought over water. In contrast, 145 water-related treaties were signed in the same period. War over water seems not to be strategically rational, hydrographically effective, or economically feasible. Shared interests along a waterway seem to consistently outweigh

---

[14] One might argue that the hot springs at Hamat Gader offer economic benefits, but these are relatively minor.

water's conflict-inducing characteristics. Furthermore, once cooperative water regimes are established through treaties, they turn out to be impressively resilient over time, even between otherwise hostile riparians and even as conflict is waged over other issues. These patterns suggest that the most valuable lesson to be learned from the history of international water disputes is that this is a resource whose characteristics tend to induce cooperation, inciting violence only as the exception.

The Jordan River basin provides a representative example in microcosm. Evidence seems to suggest that not much of the quest for negotiated boundaries has been influenced by the location of water resources. This is not to say that water has not been an important topic in each set of negotiations — quite the opposite is true. Questions about water allocations and rights have been intricate and difficult to resolve. Nevertheless, the negotiations between Israel and Jordan have been concluded, the talks between Israel and the Palestinians are ongoing, and despite the many studies identifying hydrostrategic territory and advising its retention, the location of water has not been a factor in any retention of territory. Solutions have all emphasized creative joint management.

The pattern that does seem to be emerging, however, is that, without other concerns, water does not justify retention of territory. For example, in the absence of any legal claims, security interests, or settlements, Israel withdrew from all the Jordanian territory it had occupied, even those small portions that had hydrostrategic importance. What was important was an agreement on water management, not territory.

These principles may be played out in negotiations between Israel and Syria as well. Whereas Syria insists on the Armistice Line as it stood on 5 June 1967, Israel is arguing for boundaries based on the 1923 international division between the British and French mandates, the difference being three small areas of vital hydrostrategic importance. Based on the patterns of negotiations between other coriparians in the region, once the right people are in the room and they attain a clear mandate to reach an agreement, the territorial imperative will be circumvented in favour of the principles of joint monitoring and cooperative management.

References

Alpher, J. 1994. Settlements and borders. Jaffee Center for Strategic Studies, Tel Aviv, Israel.

Beaumont, P. 1991. Transboundary water disputes in the Middle East. Paper presented at Transboundary Waters in the Middle East, Sep 1991, Ankara, AS, USA.

Bingham, G.; Wolf, A.; Wohlgenant, T. 1994. Resolving water disputes: conflict and cooperation in the U.S., the Near East, and Asia. US Agency for International Development, Washington, DC, USA. Publication ANE-0289-C-00-7044-00.

Biswas, A.K.; Hashimoto, T., ed. 1996. Asian international waters: from Ganges–Brahmaputra to Mekong. Oxford University Press, Oxford, UK.

Brecher, M.; Wilkenfeld, J. 1997. A study of crisis. University of Michigan Press, Ann Arbor, MI, USA.

Butts, K. 1997. The strategic importance of water. Parameters (spring), 65–83.

Cooley, J. 1984. The war over water. Foreign Policy, 54 (spring), 3–26.

Cooper, J. 1983. Reconstructing history from ancient inscriptions: the Lagash–Umma border conflict. Undena, Malibu, CA, USA.

Davis, U.; Maks, A.; Richardson, J. 1980. Israel's water policies. Journal of Palestine Studies, 9(2,34), 3–32.

Dillman, J. 1989. Water rights in the occupied territories. Journal of Palestine Studies, 19(1,73), pp. 46–71.

Falkenmark, M. 1986. Fresh waters as a factor in strategic policy and action. *In* Westing, A.H., ed., Global resources and international conflict: environmental factors in strategic policy and action. Oxford University Press, New York, NY, USA. pp. 85–113.

FAO (Food and Agriculture Organization of the United Nations). 1978. Systematic index of international water resources treaties, declarations, acts and cases, by basin. Vol. 1. Legislative Study 15.

———— 1984. Systematic index of international water resources treaties, declarations, acts and cases, by basin. Vol. 2. Legislative Study 34.

Fredkin, P. 1981. A river no more. Knopf, New York, NY, USA.

Gleick, P. 1993. Water and conflict: fresh water resources and international security. International Security, 18(1), 79–112.

Goldschmidt, M.; Jacobs, M. 1958. Precipitation over and replenishment of the Yargon and Nahal Taninim underground catchments. Hydrologic Service, Jerusalem, Israel.

Hamner, J.; Wolf, A. 1998. Patterns in international water resource treaties: the transboundary freshwater dispute database. Colorado Journal of International Environmental Law and Policy. 1997 Yearbook.

Homer-Dixon, T. 1994. Environmental scarcities and violent conflict. International Security (summer).

Libiszewski, S. 1995. Water disputes in the Jordan Basin region and their role in the resolution of the Arab–Israeli conflict. Center for Security Studies and Conflict Research, Zurich, Switzerland. Occasional Paper 13, Aug 1995.

Myers, N. 1993. Ultimate security: the environmental basis of political stability. Norton, New York, NY, USA.

Remans, W. 1995. Water and war. Humantäres Völkerrecht, 8(1).

Samson, P.; Charrier, B. 1997. International freshwater conflict: issues and prevention strategies. Green Cross Draft Report, May.

Schiff, Z. 1995. They are forgetting the Golan's water. Ha'aretz, 7 Jun, p. B1. [In Hebrew]

Schmida, L. 1983. Keys to control: Israel's Pursuit of Arab water resources. American Educational Trust.

Schwartz, Y.; Zohar, A. 1991. Water in the Middle East: solutions to water problems in the context of arrangements between Israel and the Arabs. Jaffee Center for Strategic Studies, Tel Aviv, Israel. [Hebrew]

Shalev, A. 1994. Israel and Syria: peace and security on the Golan Heights. Jerusalem Post Publishing, Jerusalem, Israel. Jaffee Center for Strategic Studies Study 24.

Stauffer, T. 1982. The price of peace: the spoils of war. American–Arab Affairs, 1 (summer), 43–54.

Stork, J. 1983. Water and Israel's occupation strategy. MERIP Reports 116(13,6), 19–24.

Tarnopolsky, N. 1996. Water damps hopes for deal with Syrians. Forward, 5 Jan, p. 1.

Weinberger, G. 1991. The hydrology of the Yarkon-Taninim aquifer. Hydrologic Service, Jerusalem, Israel. [Hebrew]

Westing, A.H., ed. 1986. Global resources and international conflict: environmental factors in strategic policy and action. Oxford University Press, New York, NY, USA.

Wolf, A.T. 1995a. Hydropolitics along the Jordan River: scarce water and its impact on the Arab–Israeli conflict. United Nations University Press, Tokyo, Japan.

———— 1995b. International water dispute resolution: the Middle East Multilateral Working Group on Water Resources. Water International, 20(3).

———— 1997. International water conflict resolution: lessons from comparative analysis. International Journal of Water Resources Development, 13(3).

———— 1999a. "Hydrostrategic" territory in the Jordan Basin: water, war, and Arab–Israeli peace negotiations," In Amery, H.; Wolf, A., ed., A geography of water in the Middle East at peace. University of Texas Press, Austin, TX, USA. (In press.)

———— 1999b. Water wars and water reality: conflict and cooperation along international waterways. In Lonergan, S. ed., Environmental change, adaptation, and security. Kluwer Academic Press, Dordrecht, Netherlands. (In Press.)

Conclusion

SUMMARY OF CONSENSUS FROM THE WORKSHOP PARTICIPANTS

M. Husain Sadar

Recognizing that

- Governments in the Eastern Mediterranean region are increasingly cognizant of the fundamental importance of both surface and subsurface water resources for the economic, social, and cultural well-being of their people and for development in their region;

- A large portion of humanity has strong spiritual, cultural, and historical links with the region and, as such, is very keen to preserve and protect the health and integrity of its ecosystems, each of which is fundamentally dependent on water;

- There is growing national, regional, and international understanding of the increasing threat to precious water resources from local, regional, and international sources of pollution, including, inter alia, growing tourism, chemically dependent agriculture, and globalization of trade practices;

- Comprehensive ecosystems approaches and basin-wide management strategies have proven to be the most effective ways to protect and preserve natural resources in any part of the world and, most certainly, in the Eastern Mediterranean;

- Regional cooperation agreements and joint management in their implementation are the fairest and least conflictive way to protect, preserve, and equitably share transboundary water resources, as amply demonstrated by the successful Boundary Waters Treaty of 1905 between the United States and Canada;

- Ongoing, credible, and informed advice, based on sound economic, environmental, and ecological analysis, is essential to making balanced political decisions, formulating long-term policies, and designing workable and cost-effective implementation mechanisms;

- Cooperative approaches to protecting water resources can assist governments in meeting their international commitments and obligations to protect the global ecosystem in cost-effective and efficient ways; and

- Regional cooperation, in any of its forms, should lead to the development of a feasible, focused research agenda for management and protection of watersheds, near-surface groundwater and aquifers, and their interrelated systems;

The following general conclusions and recommendations are put forward:

- All participants in the workshop are in general agreement to build further on the useful discussions held and contacts made at the conference in Ottawa;

- Because of its vast experience in transboundary watershed management and accumulated relevant knowledge and expertise in the field, Canada can and should continue to play the role of facilitator and organizer for follow-up meetings;

- It is highly desirable that a permanent Eastern Mediterranean Technical Water Advisory Group be established under the leadership of Carleton University and the International Development Research Centre (IDRC), both located in Ottawa, Canada;

- The group, once established, should consult all participants in this conference to develop a proposal and draft an agenda for the first follow-up meeting and should submit these to IDRC and Carleton University;

- One of the agenda items for the first follow-up meeting should be drafting the terms of reference for this Technical Group; and

- All researchers focusing on water issues in the Eastern Mediterranean region are urged to advise policymakers at all levels of their governments to accept, at least in principle and eventually in practice, the advantages of intra- and international cooperation for water resource management and protection.

Appendix 1

CONTRIBUTING AUTHORS

Samer Alatout
Doctoral Candidate, Science and
Technology Studies
Cornell University
Ithaca, NY, USA 14850
email: sa11@cornell.edu

Yasser Al-Adwan
Professor, Faculty of Business
Administration and Economics
Yarmouk University
Irbid, Jordan 21163
email: adwany@yu.edu.jo

Hussein A. Amery
Assistant Professor, Division of Liberal
Arts and International Studies
Colorado School of Mines
Golden, CO, USA 80401
email: hamery@mines.edu

Hasan Ali Biçak
Associate Professor, Department of
Banking and Finance
Eastern Mediterranean University
Gazimagusa, Mersin 10, Turkey
email: bicak@salamis.emu.edu.tr

David B. Brooks
Research Manager, Programs Branch
International Development Research
Centre
PO Box 8500
Ottawa, ON, Canada K1G 3H9
email: dbrooks@idrc.ca

Glenn P. Jenkins
Institute Fellow, Harvard Institute for
International Development
1 Eliot Street
Cambridge, MA, USA 02138
email: gjenkins@hiid.harvard.edu

Harvey Lithwick
Professor, Negev Center for Regional
Development
Ben-Gurion University of the Negev
Beer-Sheva, Israel 84105
email: lithwick@gumail.bgu.ac.il

Ozay Mehmet
Professor, Norman Paterson School of
International Affairs
Carleton University
Ottawa, ON, Canada K1S 5B6
email: omehmet@ccs.carleton.ca

M. Husain Sadar
Adjunct Professor of Environmental
Science, School of Natural Sciences
Carleton University

Ottawa, ON, Canada K1S 5B6
email: husain_sadar@carleton.ca

Esam Shannag
Chair, Biology Department
Yarmouk University
Irbid, Jordan 21163
email: eshannag@yu.edu.jo

Mehmet Tomanbay
Professor
Gazi University
Ankara, Turkey
email: mtomanbay@yahoo.com

Aaron T. Wolf
Assistant   Professor,   Department   of
Geosciences
Oregon State University
Corvallis, OR, USA 97331
email: wolfa@geo.orst.edu

Appendix 2

ACRONYMS AND ABBREVIATIONS

BOT	Build–Operate–Transfer
dn	donum [1 dn = 0.1338 ha]
MPCC	Multi-purpose Community Centre
GAP	Guneydogu Anadolu Projesi (Southeast Anatolia Project)
GAP ESGC	Southeast Anatolia Project Entrepreneur Support and Guidance Centres
GDP	gross domestic product
GNP	gross national product
IDRC	International Development Research Centre
JWC	Joint Water Committee
MWR	minimum water requirement
NGO	nongovernmental organization
NPV	net present value
OID	Organized Industrial District
SCBA	social cost–benefit analysis
SIE	Small Industrial Estate
STS	science and technology studies
TRL	Turkish lira
USD	United States dollar

About the Institution

The International Development Research Centre (IDRC) is committed to building a sustainable and equitable world. IDRC funds developing-world researchers, thus enabling the people of the South to find their own solutions to their own problems. IDRC also maintains information networks and forges linkages that allow Canadians and their developing-world partners to benefit equally from a global sharing of knowledge. Through its actions, IDRC is helping others to help themselves.

About the Publisher

IDRC Books publishes research results and scholarly studies on global and regional issues related to sustainable and equitable development. As a specialist in development literature, IDRC Books contributes to the body of knowledge on these issues to further the cause of global understanding and equity. IDRC publications are sold through its head office in Ottawa, Canada, as well as by IDRC's agents and distributors around the world. The full catalogue is available at http://www.idrc.ca/books/index.html.